ENVIRONMENTAL PERCEPTION AND BEHAVIOR

An Inventory and Prospect

Thomas F. Saarinen, David Seamon, and James L. Sell
Editors

James M. Blaut
Martyn J. Bowden
Anne Buttimer
Helen Couclelis
Robin E. Datel
Dennis J. Dingemans
Larry R. Ford
John R. Gold
Reginald G. Golledge

Douglas Greenberg
Roger Hart
James K. Mitchell
Robert Mugerauer
Edward Relph
Gerard Rushton
Joseph Sonnenfeld
Jonathan G. Taylor
Gilbert F. White

Ervin H. Zube

THE UNIVERSITY OF CHICAGO
DEPARTMENT OF GEOGRAPHY
RESEARCH PAPER NO. 209

1984

Copyright 1984
by
Thomas F. Saarinen, David Seamon, and James L. Sell
Published 1984
by
The Department of Geography
The University of Chicago
Chicago, Illinois

Library of Congress Cataloging in Publication Data
Main entry under title:

Environmental perception and behavior.

  (Research papers / University of Chicago, Dept. of Geography ; 209)
  Includes bibliographical references.
  1. Geographical perception. I. Saarinen, Thomas F. (Thomas Frederick) II. Seamon, David. III. Sell, James L., 1947-   . IV. Series: Research paper (University of Chicago. Dept. of Geography) ; no. 209.
H31.C514 no. 209   [G71.5]   910s [912'.01'4]   84-2492
ISBN 0-89065-114-0 (pbk.)

Research Papers are available from:
The University of Chicago
The Department of Geography
5828 S. University Avenue
Chicagom Illinois   60637
Price: $8.00; $6.00 by series subscription

CONTENTS

LIST OF TABLES v

LIST OF ILLUSTRATIONS vii

A NOTE ON THE TEXT AND THE CONTRIBUTORS ix

INTRODUCTION
     Thomas F. Saarinen, David Seamon, and James L. Sell . . 3

PART I
REVIEWS OF RESEARCH
GENERAL AND LARGE-SCALE THEMES

Chapter
1. SOME REASONS FOR OPTIMISM ABOUT ENVIRONMENTAL PERCEPTION RESEARCH
   Thomas F. Saarinen 13

2. BEHAVIORAL GEOGRAPHY IN WESTERN EUROPE: REFLECTIONS ON RESEARCH IN GREAT BRITAIN AND THE FRANCOPHONE NATIONS
   John R. Gold 25

3. HAZARD PERCEPTION STUDIES: CONVERGENT CONCERNS AND DIVERGENT APPROACHES DURING THE PAST DECADE
   James K. Mitchell 33

4. TOWARD A THEORETICAL FRAMEWORK FOR LANDSCAPE PERCEPTION
   James L. Sell, Jonathan G. Taylor, and Ervin H. Zube . 61

5. ENVIRONMENTAL PERCEPTION IN GEOGRAPHY: A COMMENTARY
   Martyn J. Bowden 85

6. ENVIRONMENTAL PERCEPTION AND ITS USES: A COMMENTARY
   Gilbert F. White 93

PART II
REVIEWS OF RESEARCH
URBAN AND MICROSCALE THEMES

7. THE GEOGRAPHY OF CHILDREN AND CHILDREN'S GEOGRAPHIES
   Roger Hart 99

8. ENVIRONMENTAL PERCEPTION, HISTORIC PRESERVATION, AND SENSE OF PLACE
   Robin E. Datel and Dennis J. Dingemans 131

9. WHERE DO WE GO FROM HERE? A COMMENTARY
   Larry R. Ford 145

10. MODESTY AND THE MOVEMENT: A COMMENTARY
    James M. Blaut 149

PART III
PHILOSOPHICAL DIRECTIONS

11. PHILOSOPHICAL DIRECTION IN BEHAVIORAL GEOGRAPHY WITH AN EMPHASIS ON THE PHENOMENOLOGICAL CONTRIBUTION
    David Seamon 167

12. POSITIVIST PHILOSOPHY AND RESEARCH ON HUMAN SPATIAL BEHAVIOR
    Reginald G. Golledge and Helen Couclelis 179

13. WHODUNIT? STRUCTURE AND SUBJECTIVITY IN BEHAVIORAL GEOGRAPHY
    Douglas Greenberg 191

14. SEEING, THINKING, AND DESCRIBING LANDSCAPES
    Edward Relph 209

15. PHILOSOPHICAL DIRECTIONS IN ENVIRONMENTAL PERCEPTION AND BEHAVIORAL GEOGRAPHY: A COMMENTARY
    Joseph Sonnenfeld 225

16. MAPPING THE MOVEMENT OF GEOGRAPHICAL INQUIRY: A COMMENTARY
    Robert Mugerauer 235

17. MAINTAINING THE FUNDAMENTAL PRINCIPLES: A COMMENTARY
    Gerard Rushton 245

18. PERCEPTION IN FOUR KEYS: A COMMENTARY
    Anne Buttimer 251

LIST OF TABLES

1.1	Perception Themes in Recent Geography Dissertations	16
1.2	Intellectual Origins of Perception of Environment Group	18
4.1	Evaluation of Paradigms	69
4.2	Paradigm Contributions to a Transactional Approach to Perception of Environment	73
5.1	Behavioral Geography	86

LIST OF ILLUSTRATIONS

3.1	Human Adjustments to Natural Hazards	34
3.2	Key Concepts in the Emerging Theory of Risk Mitigating Adjustment	35
3.3	Causes and Consequences of Risk Perception	50
4.1	The Emergence of Landscape Perception Research Paradigms	62
4.2	Landscape Perception Interaction Process	70
4.3	Integrated Framework for Landscape Perception Research	82
5.1	Periods of Vitality in Cognitive/Perceptual Studies in Geography	88

A NOTE ON THE TEXT AND THE CONTRIBUTORS

The contributions to this volume are edited versions of papers presented at the annual meeting of the Association of American Geographers, San Antonio, Texas, April 28, 1982. They represent the work of two special sessions sponsored by the Environmental Preception Specialty Group of the AAG. One session, actually a double session, titled "Environmental Perception: An Inventory and Prospect," was organized and chaired by Thomas F. Saarinen and James L. Sell. The other, titled "Philosophical Directions in Environmental Perception and Behavioral Geography," was organized and chaired by David Seamon. The names and affiliations of the contributors are presented below in alphabetical order.

James M. Blaut
Department of Geography
University of Illinois
Chicago Circle

Martyn J. Bowden
Graduate School of Geography
Clark University

Anne Buttimer
Department of Social and
  Economic Geography
University of Lund

Helen Couclelis
Department of Geography
University of California
Santa Barbara

Robin E. Datel
Department of Geography
University of Minnesota

Dennis J. Dingemans
Department of Geography
University of California, Davis

Larry R. Ford
Department of Geography
San Diego State University

John R. Gold
Department of Geography
Oxford Polytechnic University

Reginald G. Golledge
Department of Geography
University of California
Santa Barbara

Douglas Greenberg
Department of Geography
University of California, Berkeley

Roger Hart
Environmental Psychology Program
City University of New York

James K. Mitchell
Department of Environmental
  Resources
Rutgers University

Robert Mugerauer
Department of Philosophy
St. Edward's University

Edward Relph
Department of Geography
University of Toronto

Gerard Rushton
Department of Geography
University of Iowa

Thomas F. Saarinen
Department of Geography
University of Arizona

David Seamon
Department of Architecture
Kansas State University

James L. Sell
Department of Geography
University of Arizona

Joseph Sonnenfeld
Department of Geography
Texas A & M University

Jonathan G. Taylor
Office of Arid Land Studies
University of Arizona

Gilbert F. White
Institute of Behavioral Science
University of Colorado

Ervin H. Zube
School of Renewable Natural Resources
University of Arizona

ENVIRONMENTAL PERCEPTION AND BEHAVIOR
AN INVENTORY AND PROSPECT

INTRODUCTION

Thomas F. Saarinen, David Seamon, and James L. Sell

The first formal session on environmental perception and behavior to be presented at an Association of American Geographers annual convention took place at Columbus, Ohio, on April 20th, 1965. The papers presented later appeared as a volume in the University of Chicago Department of Geography Research Series, which provides a benchmark for comparison with the papers in the present volume.[1] These were also based on AAG sessions organized for the San Antonio meeting in 1982, to provide an inventory and prospect of the same field. Both environmental perception research and geography have changed considerably in the intervening seventeen years. Comparison of this volume with the earlier one edited by Lowenthal clearly illustrates some of the major changes.

That early volume contained five papers by a highly motivated group of pioneers investigating ways to approach the problem of understanding the connections between perception, behavior, and environment. These approaches included: the use of literature and art in understanding attitudes toward environment,[2] the development of a psychological test for analyzing spatial symbolism,[3] the incorporation of psychological theory and methodology into an examination of perception and adaptation in the Arctic,[4] the employment of survey research in the study of attitudes to storm

---

1. David Lowenthal, ed., *Environmental Perception and Behavior*, Research Paper no. 109, (Chicago: University of Chicago, Department of Geography, 1967). The title used contains four words and therefore is too long for common usage as the name of the field. Environmental perception or perception of environment have been the terms most generally used to refer to the entire field ever since. Here, we often use the better term, behavioral geography, interchangeably, although in North American geography behavioral geography has most commonly been used to refer to the theoretical, quantitative, model-building branch of the field, closely aligned to the spatial tradition in geography, and represented in this volume by such authors as Golledge, Couclelis, and Rushton.

2. Yi-Fu Tuan, "Attitudes toward Environment: Themes and Approaches," ibid., pp. 4-17.

3. Robert Beck, "Spatial Meaning, and the Properties of the Environment," ibid., pp. 18-41.

4. Joseph Sonnenfeld, "Environmental Perception and Adaptation Level in the Arctic," ibid., pp. 42-59.

hazards,[5] and the combination of image and experiential view in the perception of landscapes as viewed from highways.[6]

Since that groundbreaking session, the interest in environmental perception and behavior has grown tremendously, as can be seen in the reviews and discussions in this volume, which represent a large portion--but not all--of the concerns of the field.[7] The current volume presents the results of the Environmental Perception Specialty Group's first attempt to organize a set of sessions for an AAG meeting. The enthusiastic reactions by the contributors and by those attending the sessions, as well as the interest shown by those requesting information later, encouraged us to collect at least some of the presentations in volume form. The result is this collection, which assembles papers from three of the sessions: Saarinen and Sell's double session on "Environmental Perception: Inventory and Prospect;" and Seamon's "Philosophical Directions in Environmental Perception and Behavioral Geography."[8]

Part I includes general assessments of the field in North America and Europe, as well as the large-scale themes of environmental hazards and landscape perception. In his review of the field, Saarinen takes a decidedly optimistic view pointing to growth in the quality and quantity of work, the institutionalization of research, and the integration with the mainstream of geography, as signs of the maturity of environmental perception and behavior as a subdiscipline. Gold's view from Europe is more pessimistic. There, the burgeoning growth of Francophone interest is contrasted with Great Britain, where behavioral geography is at a low ebb. Hazards Research, as outlined by Mitchell, is a large body of work in natural, technological, and social hazards, in which hazard perception is a small, somewhat controversial, proportion. Landscape perception is seen by Sell, Taylor, and Zube to encompass

---

5. Robert W. Kates, "The Perception of Storm Hazard on the Shores of Megalopolis," ibid., pp. 60-74.

6. Donald Appleyard, Kevin Lynch, and John R. Meyers, "The View From the Road," ibid., pp. 75-88.

7. We originally invited individuals to prepare papers on cognitive mapping, and behavioral geography (the theoretical, quantitative, model-building branch noted in footnote 1) for our inventory and prospects sessions. These were the only invited papers not completed. Hart's paper on children highlights an important trend toward more sensitive appraisal of person-environment relationships by careful examination of major population subgroups. Much research has also appeared on the elderly, the handicapped, and women.

8. Because of space requirements, papers from Marilyn Brown's special session, "Recent Emirical Research in Behavioral Geography," could not be included. Participants in that session were G. Donald Richardson ("Understanding Space: the Case of the Mentally Retarded"); Pamela Bergmann and K. David Pijawka ("Effects of the Three Mile Island Accident on Residents Residing Near Nuclear Power Stations"); and Marilyn Brown ("Attitude Change and the Adoption of Residential Energy Conservation Measures"). Discussants for the session were Michael Dear, Valerie Preston, and Susan Caris Cutter.

a number of diverse interdisciplinary approaches in which geographers have played only a small role. Two commentaries complete the first part. Bowden's comments examine the success of the "cognitive renaissance" of the 1960s in light of the failures of the 20s and 40s in North America, and of the present in Great Britain. White also provides perspective in discussing the reasons for organizing the Columbus session. He suggests that a current assessment of the field should include an examination of the degree to which ideas are applied in practical planning contexts, as well as systematic comparisons of the methods and techniques already tried by various researchers.

Part II of the volume reviews themes relating to micro-scale and urban levels of research, including children's environmental experience, and historic preservation and sense of place. Hart laments the lack of geographic research on the geography of children, and children's geographies, despite their unequivocally geographic nature. Datel and Dingemans discuss the interrelationships among sense of place, historic preservation and environmental perception research. In the two commentaries concluding Part II, Ford asserts that the importance of environmental perception research involves its integration with traditional areas of study in geography, while Blaut explains many sources of error in early cross-cultural "psychology," as he outlines various "modest" and "immodest" approaches.

Essays in Part III move beyond discussion of existing research and examine wider philosophical issues. Organized around the outline provided by Derek Gregory's *Ideology, Science and Human Geography,* these works focus on positivist, reflexive, and structuralist interpretations of environmental perception, behavior and experience.[9] Seamon looks at the phenomenological contribution to research in environmental behavior and experience, while Golledge and Couclelis examine positivist philosophy and its changes in light of research in behavioral geography. Yet again, Greenberg presents the Marxist-structuralist interpretation of ideology and individual consciousness, while Relph outlines the value of reflexive, personal approaches in promoting a better understanding of landscapes and places. In the four commentaries of Part III, Sonnenfeld and Rushton both discuss in general terms the strengths and weaknesses of the dominant positivist approach in environmental perception and behavioral geography. Buttimer turns to the sociology of the

---

9. Derek Gregory, *Ideology, Science and Human Geography* (London: Hutchinson, 1978).

movement and speaks of the four perspectives of positivism, Marxism, structuralism, and phenomenology as imports from European thought adapted to North American theory and practice. Finally, Mugerauer, a philosopher, points to the importance of the classical dichotomy of "subjectivity" versus "objectivity" as a key means for understanding the various philosophical styles and methods currently affecting trends in environmental perception and behavioral geography.

Clearly, the essays of this volume indicate that the field of environmental perception and behavior is considerably more complex than it was in 1967. Complications arising from different philosophical positions are evident, many methods are used, and the array of substantive studies is diverse. Some of this diversity may be attributed to the rapid acceptance of the behavioral approach by almost all segments of the field of geography. This rapid acceptance may be illustrated by examination of AAG presidential addresses immediately following the national meeting at Columbus, Ohio. The addresses by James,[10] Burrill[11] and Kollmorgen[12] could be considered mainly behavioral in orientation, while Kohn spoke of environmental perception as a frontier area of geographic research in his address on "The 1960s: A Decade of Progress in Geographical Research and Instruction."[13] The first post-Columbus presidential address which failed to incorporate references to the behavioral approach was that on "The World of Underground Ice" by Mackay in 1971.[14] This rapid acceptance may be attributed to the compatibility of the ideas and concepts of environmental perception with many older geographic traditions. Indeed the presidential addresses of Barrows,[15] Marbut,[16] Whittlesey,[17] Wright,[18] and Leighly,[19] included

---

10. Preston E. James, "On the Origin and Persistence of Error in Geography," *Annals of the Association of American Geographers* 57 (1967): 1-24.

11. Meredith F. Burrill, "The Language of Geography," *Annals AAG* 58 (1968): 1-11.

12. Walter M. Kollmorgan, "The Woodman's Assault on the Domain of the Cattleman," *Annals AAG* 59 (1969): 215-239.

13. Clyde F. Kohn, "The 1960s: A Decade of Progress in Geographical Research and Instruction," *Annals AAG* 60 (1970): 211-219.

14. J. Ross Mackay, "The World of Underground Ice," *Annals AAG* 62 (1972): 1-22.

15. Harlan H. Barrows, "Geography as Human Ecology," *Annals AAG* 13 (1923): 1-14.

16. Curtis F. Marbut, "The Rise, Decline, and Revival of Malthusianism in Relation to Geography and the Character of Soils," *Annals AAG* 15 (1925): 1-29.

17. Derwent Whittlesey, "The Horizon of Geography," *Annals AAG* 35 (1945): 1-36.

18. John K. Wright, "Terrae Incognitae: The Place of Imagination in Geography," *Annals AAG* 37 (1947): 1-15.

such themes prior to the most recent emergence of environmental perception studies. As Bowden states in his commentary, the 60s were really not the beginning but the third period of vitality in cognitive-perceptual studies in geography.

Other indicators of the widespread acceptance of the behavioral approach may be seen in the chapter by Saarinen, particularly in discussion of the intellectual antecedents of the earliest group to join the environmental perception specialty group. What is generally not appreciated by geographers not directly involved, is the diversity of intellectual origins of the several distinct lines of research which together made up the field known as "perception of environment." This diversity is at once a major strength and a source of difficulty. In their commentaries, both Bowden and Buttimer note the strong integration of the many approaches at the onset of perception research in the 1960s and see some fragmentation into separate research directions as a potential source of weakness. According to Gold, this potential problem became real in Great Britain. There, the dispersion of individuals sapped the vitality of a promising start by a small interdisciplinary group.

Dispersion of individuals into an enormous number of subgroups has also taken place in North American behavioral geography. There is diversity, not only in topics of study but in basic philosophical premises, as is noted above and in many of the chapters which follow. Greenberg, for example, speaks of the two major approaches to behavior-perception research as scientific-positivist, and phenomenological-humanist. Each of these could be further subdivided. Buttimer identifies four separate schools of thought, and Bowden separates out five distinct facets in his consideration of the first four papers. Further centrifugal forces are created by the strong ties of various individuals or subgroups to the many neighboring fields in the social sciences, humanities, and design disciplines.

The focus on practical planning problems has also contributed to the diversity within behavioral geography. White notes that much of the impetus for the development of the behavioral approach in geography and behind the organization of the original session at Columbus, was the desire to find methods useful for dealing with practical human problems. Many researchers proceeded with the task unhindered by any strong resistance to their methods and ideas

---

[19]. John B. Leighly, "John Muir's Image of the West," *Annals AAG* 48 (1958): 309-318.

within the geographic profession. This development is in sharp contrast to the quantitative revolution which faced strong resistance because of unfamiliar models and methods. The summer institutes organized to teach, develop, and standardize models and methods helped the quantifiers to develop the evangelical fervor and *esprit de corps* needed to establish themselves. An equivalent central focus never developed for those interested in environmental perception and behavior. In spite of, or perhaps, because of this eclecticism, new methods, ideas and concepts continue to be developed, and old ones continue to be applied in many new contexts.

A consequence of the focus on practical problems may in certain cases be the type of outcome described by Mitchell in relation to hazards research. Here, the proportion of perception studies to total studies shifted over time, from none to a position of dominance, then to the current situation of a rather modest percentage. Interestingly, as this pattern occurred in relation to natural hazards, the behavioral frontiers have been advanced by some researchers to focus on such derivatives as technological hazards, risk assessment, and problems of human rationality. As the frontiers advance, new methods are added to overcome new deficiencies which have been perceived by the researchers or pointed out by their critics. Often, corrections are made before criticisms appear.[20] The vitality exemplified is vitiated to some degree by the dispersal of energies to an ever-advancing series of problems. Only incidentally do individual researchers attempt to assess the methods used by the entire field.

A major function of the Environmental Perception Speciality Group is to help provide more focus for the integration of these diverse points of view. This focus is seen as important by the members of our group and it was for this reason that the special sessions were organized for the meetings in San Antonio. The felt need for more integration is apparent in many of the chapters, such as the commentary by Ford. He shows how the various environmental perception studies dovetail with four long-established traditions of geography. Appreciation of this pluralistic approach, with its different and even sometimes strongly contrasting viewpoints, can help to focus rather than disperse energies. The topic of sense of place reviewed in the Datel and Dingemans chapter has potential as one means of integrating the various geographic traditions. The chapters by Hart, and Sell, Zube, and Taylor represent determined

---

20. Robert W. Kates, "Hazard Research: A View From Worcester," a paper prepared for the Cultural Ecology Speciality Group, Association of American Geographers Annual Meeting, Denver, Colorado, April, 1983.

efforts to integrate the useful research results of several contrasting approaches to the solution of particular practical concerns.

Similarly, one can see within the philosophical set of papers a certain conciliatory spirit. Though strong exponents of the positivistic approach, Golledge and Couclelis acknowledge that other approaches may be more appropriate for some behavioral problems. In the same way, Mugerauer indicates that much agreement is apparent in spite of differing philosophical stances. In the end, it is not surprising that the session in which the exponents of various philosophical persuasions presented the advantages of their position took place under the auspices of the Environmental Perception Specialty Group. A major conviction motivating such research is that there are many ways to interpret the environment, and human planning should make an effort to understand and, if possible, incorporate the views of those people who will be affected.

We remain optimistic about the future of behavioral research in geography in spite of the rather dispersed nature of current efforts. The increasing number of researchers, the tremendous vitality that has been generated by the interdisciplinary interaction of the past couple decades, and the impetus likely to be produced by the recent international developments, promise an exciting future for research and application of environmental perception and behavior. Although optimistic, we feel that more focussed discussion of past findings and future plans are needed to maximize our efforts. We hope that this volume will not only illustrate the diversity of approaches but will also help to provide that focus.

PART I
REVIEWS OF RESEARCH
GENERAL AND LARGE-SCALE THEMES

CHAPTER 1
SOME REASONS FOR OPTIMISM ABOUT PERCEPTION OF ENVIRONMENT RESEARCH

Thomas F. Saarinen

The optimistic outlook on perception of environment research which I hold is based on my experience in attempting to comprehensively review the literature of the field. More precisely, it resides in the dramatic contrast I see between the amount and quality of research available in the early 1960's and that found today.

My most recent attempts to review the perception of environment research began with a request from the editors of *Progress in Human Geography* to prepare for them three progress reports on the field, one each year for three consecutive years. I enlisted the aid of Jim Sell. We have now completed the three reviews (the final one with the additional aid of Eliza Husband).[1]

Prior to the reviews for *Progress in Human Geography* I had attempted to assess the entire field of environmental perception about 5 years earlier.[2] I expected a great contrast between the literature in the early 60s, when the field was in its infancy, and today. What I was not entirely prepared for was the enormous growth and development of the past five years. The growth pattern is a geometric progression resulting in a flood of information. By 1978 Wheeler stated, "the publication rate, now, implies that if you were to read one *new* article or book on environmental psychology, *every single day of the year,* you probably would not be able to keep up with the literature!"[3] Today it is even more difficult.

Numbers of Geographers Specializing in Environmental Perception and Impact on the Field of Geography

A quantitative measure of the increasing interest in environmental perception in North America is provided by the Association of American Geographers (AAG) members who list it as one

---

1. Thomas F. Saarinen and James L. Sell, "Environmental Perception," *Progress in Human Geography,* 4, (1980): 535-548; ibid., 5 (1981): 525-547; and Thomas F. Saarinen, James L. Sell, and Eliza Husband, "Environmental Perception: International Efforts," *Progress in Human Geography,* 6 (1982): 515-546.

2. Thomas F. Saarinen, *Environmental Planning: Perception and Behavior,* (Boston: Houghton-Mifflin, 1976).

3. Lawrence Wheeler, "An ABC History of Environmental Psychology," unpublished paper, University of Arizona, 1978.

of their topical specialities on the AAG membership forms. From a base of close to zero in 1960 this grew to 193 by 1971, 234 by 1975, and 310 by 1979.[4] Thus from 1971 to 1975 the percentage increase was 21.2. This rose to 32.5 for the 1975-79 period. As a proportion of the total AAG membership, the group professing environmental perception as a topical proficiency grew from 2.9% in 1971, to 3.3% in 1975, and 5.3% in 1979. These certainly appear to be signs of healthy growth but I believe they underestimate the impact of the field of environmental perception on the geographic profession.

The 1978 Directory of the Association of American Geographers was used to obtain the names and addresses of the geographers who listed environmental perception as a topical specialty.[5] We felt a direct mail request for reprints, preprints, and references would be the most efficient way to sample the current literature because of the wide dispersion of environmental perception articles in academic journals of all types. To this original list we added any additional names and addresses obtainable in the 1979-80 AAG Guide to Graduate Departments in the United States and Canada.[6] In addition, we added the names of British geographers we knew had published in this field. This provided us with a list of several hundred geographers. At a later stage we also added the names of many people in other disciplines.

To extend our mailing list we looked at the program for the AAG meeting in Philadelphia[7] and later at those for the AAG meetings in Louisville[8] and in Los Angeles.[9] In so doing we obtained strong evidence of the degree to which the geography profession has accepted and adopted the ideas and techniques of environmental perception research. In each case a large number of papers were presented which could be classified as falling within the field of environmental perception by virtue of their focus on behavior, preferences, attitudes, images, decision-making, or use of questionnaire surveys, cognitive maps or other behavioral methodologies (in Los Angeles close to 1/5 of the papers could be so classified).

---

4. "Proficiency Codes," *AAG Newsletter* 15 (1980): 16.

5. AAG, *Directory of the Association of American Geographers* (Washington, D.C.: Association of American Geographers, 1978).

6. AAG, *Guide to Graduate Departments of of Geography in the United States and Canada 1980-81* (Washington, D.C.: Association of American Geographers, 1979).

7. AAG, *Program Annual Meeting April 22-25, 1979, Philadelphia, Pennsylvania*.

8. AAG, *Program Annual Meeting April 12-16, 1980, Louisville, Kentucky*.

9. AAG, *Program Annual Meeting April 19-22, 1981, Los Angeles, California*.

Even more striking was comparison of the names of those presenting papers on environmental perception themes at AAG national meetings with the list of geographers who indicated environmental perception as a topical speciality. Over half of those presenting papers on environmental perception themes at the Philadelphia AAG meetings were not on our list. We added their names. Comparing this new extended list with the names of people presenting papers on environmental perception at the Louisville meetings we again found that over half of the authors of these papers were not on the list. The new names from the Louisville AAG program were added to our list. The names of those presenting papers on environmental perception themes at the Los Angeles AAG meeting was compared with this further extended list. Once again well over half of the people presenting papers on environmental perception were not on our list.

This exercise undertaken simply to extend our mailing list indicated to us that environmental perception themes are not simply used by a small group of geographers specializing in that field. Instead they are widely diffused throughout the geography profession. I would contend that this is so because the ideas, concepts and techniques of environmental perception have been widely accepted as useful, important, or interesting and seen as applicable in many different contexts. They have become thoroughly integrated into geographic research just as the wave of quantitative techniques which preceded them.

Further evidence of the growing influence of environmental perception themes in geography may be seen in table 1.1 which indicates the proportion of recent M.A. theses and Ph.D. dissertations that focus on such themes. Clearly the proportion of newly graduating geographers who are utilizing environmental perception themes (close to 14%) is much greater than the proportion of the profession who included it as a topical speciality (5.3% in 1979). Such themes are used even in physical geography but are most common in social geography where over a fifth could be so classified. If one assumes that the new graduates tend to be among the most active in the profession and that they continue to do research related to their dissertation topics, then one might expect a continued high level of production of research on environmental perception themes within the profession of geography.

Another measure of the grass roots acceptance of environmental perception in geography may be seen in the results of a preliminary survey we carried out to help us decide whether to try to organize an environmental perception specialty group. The first wave of

TABLE 1.1

PERCEPTION THEMES IN RECENT GEOGRAPHY DISSERTATIONS

Source	Number with Perception Themes[b]	Total Number	% with Perception Themes
A. AAG Guide to Graduate Departments[a]			
PhD dissertations USA	20	129	15.5
PhD dissertations Canada	2	29	6.9
Master's degrees USA	33	211	15.6
Master's degrees Canada	20	153	13.1
Total all PhD and Master's Theses	75	522	14.4
B. University Microfilm International[c]			
Physical Geography	2	76	2.6
Social Geography	39	180	21.6
Transportation	7	71	9.9
Urban and Regional Planning	24	189	12.7
Total	72	516	14.0

[a]*Guide to Graduate Departments of Geography in the United States and Canada 1980-1981* (Washington, D.C.: Association of American Geographers, 1980).

[b]Perception themes were defined by titles which indicated a focus on behavior, perception, attitudes, preferences, decision-making, or use of questionnaire, surveys, cognitive maps, or other behavioral methodologies as the main thrust of the study.

[c]Taken from the list of University Microfilms International, "Geography: A Dissertation Catalog." The degree dates included in the catalog range from 1978 through 1980. The list includes some dissertations from fields other than geography.

replies, which came in time for analysis prior to the AAG meetings in Los Angeles, consisted of 73 persons who were willing to join an environmental perception specialty group.

To gain some idea of the intellectual antecedents of this group, I used the AAG directory[10] to determine the department in which they received their highest degree. Thirty-six different departments were represented from four different countries, the United States, Great Britain, Canada, and France (see table 1.2). Only two departments had enough graduates represented to make up as much as ten percent of the group. Not surprisingly these were Chicago, where the early research on perception of natural hazards began, and Clark, where an early and intensive interaction between psychologists and geographers took place. Included in both cases were graduates from several time periods rather than a single group of years. To me, the most interesting thing was the enormous variety of departments represented. It is in keeping with an impression I have gained in visiting a large number of geography departments over the years. Often there are students very interested in the ideas and concepts of environmental perception even at universities where there are no faculty members who profess to specialize in the topic. In other words, many of these researchers were self-taught. In the 60s and early 70s, such students would have to seek out information on perception of environment in the research literature. Now it is easier. One finds discussion of behavioral geography, by whatever name, in most good introductory human geography texts,[11] and more and more often in texts of other more specialized fields,[12] including a growing number on perception of environment.[13] A survey of British geography departments by Gold revealed that behavioral geography is widely taught there.[14] Eighty-five percent of the departments replied to

---

10. AAG, *Directory*, 1978.

11. The first of these discussions appeared as a small section in Jan O.M. Broek and John W. Webb, *A Geography of Mankind* (New York: McGraw-Hill, 1968). Extensive discussions of behavioral factors are found in four separate chapters of Peter Haggett, *Geography: A Modern Synthesis* (New York: Harper and Row, 1975).

12. An early example in political geography was Roger E. Kasperson and Julian V. Minghi, *The Structure of Politicial Geography* (Chicago: Aldine, 1969). Three recent urban geography textbooks with special emphasis on behavioral approaches are Risa Palm, *The Geography of American Cities* (New York: Oxford University Press, 1981); Truman A. Hartshorn, *Interpreting the City: An Urban Geography* (New York: John Wiley & Sons, 1980), and David Ley, *A Social Geography of the City* (New York: Harper and Row, 1983).

13. The earliest of these was more like a literature review while the most recent resembles a textbook. See Thomas F. Saarinen, *Environmental Planning: Perception and Behavior* (Boston: Houghton-Mifflin, 1976); J. Douglas Porteous, *Environment and Behavior: Planning and Everyday Urban Life* (Reading, Mass: Addison-Wesley, 1977), and John R. Gold, *An Introduction to Behavioural Geography* (Oxford: Oxford University Press, 1980).

14. John R. Gold, "Teaching Behavioral Geography," *Journal of Geography in Higher Education* 1 (1978): 37-46.

TABLE 1.2

INTELLECTUAL ORIGINS OF PERCEPTION OF ENVIRONMENT GROUP
SPRING 1980 (n = 73)

Institution of Highest Degree*	Number	%
Chicago	8	11
Clark	7	10
Minnesota	4	5
Berkeley	3	4
Illinois	3	4
Iowa	3	4
Johns Hopkins	3	4
McMasters	3	4
University of California, Santa Barbara	2	3
Hull	2	3
Maryland	2	3
Michigan	2	3
Northwestern	2	3
Ohio State	2	3
Pittsburgh	2	3
Southern Illinois	2	3
Syracuse	2	3
Washington	2	3
Wisconsin	2	3

*Seventeen departments represented by one person were Bristol, Calgary, California, Catholic University of America, Edinburgh, Grenoble, Harvard, Indiana, Kentucky, McGill, Michigan State, Nebraska, North Dakota, Oregon, Pennsylvania State, Rutgers, and Toronto.

his mail survey and of these only one indicated that no component of behavioral geography was taught. A similar pattern seems to be emerging from the data on American geography departments currently being analyzed by Good.[15] All these changes are illustrations of the institutionalization of the field.

---

15. James K. Good, personal communication, January 28, 1982.

Institutionalization of the Field

In the previous section, the discussion focussed mainly on recent signs of vitality of environmental perception research in North American geography. It is important to bear in mind that the field I have been speaking about is really an interdisciplinary one and that to speak of it only in the context of one discipline is misleading. The geographic research is part of an advancing research frontier at the nexus of environment-behavior-design which spans a great variety of disciplines. Throughout this entire range of disciplines, from the social sciences to the design professions, there is evidence of a growing institutionalization of the field and of the various subareas.

Institutionalization of the environment-behavior-design field may be seen in the growing number of trained researchers, specialized research institutions, specialized networks of researchers, associations, publication outlets, regular meetings, and funding possibilities, which have developed. A notable trend has been a greater integration of the various disciplines involved.

Natural hazards research, one of the original subfields of environmental perception in geography, provides a good example of the type of institutionalization which has developed and may be expected in other subfields as they continue to grow and develop.[16] An increase in the number of trained researchers in the hazards field was made possible by the provision of graduate training in a number of separate universities. During the 1970s the two major social science groups, the sociologists concerned with disaster studies and the geographers focussing on natural hazards, became more thoroughly integrated. This was accomplished by an assessment of the broad base of empirical studies from both fields by a geographer and a sociologist.[17]

The Natural Hazards Research and Applications Information Center (NHRAIC) was established at the University of Colorado in 1976. It plays a clearing-house role, to facilitate the exchange between producers and users of information on natural hazards. This is done by means of a quarterly newsletter, *The Natural Hazards Observer,* an annual national workshop as well as other irregularly scheduled workshops on special topics, a monograph and discussion paper series, and through the use of its library collection and information services by those interested in natural hazards or disaster information. A similar role is played by the Disaster Research Center (DRC) at Ohio State University.

---

16. William A. Anderson, "Social Science Disaster Research in the United States," *Emergency Planning Digest* (1979): 20-24.

17. Gilbert F. White and J. Eugene Haas, *Assessment of Research on Natural Hazards* (Cambridge, Mass: The MIT Press, 1975).

Researchers have developed close contacts with appropriate government agency personnel and greater funding possibilities are now available through specific programs focussing on hazards. The recent formation of the Federal Emergency Management Agency which now houses most of the federal level emergency management concerns could have the long-term effect of further strengthening the role of social science research on natural hazards.

Similar developments have occurred in other subfields of environment-behavior-design research. The final report of the American Psychological Association's Task Force on Environment and Behavior serves as a resource book for information on the literature, graduate programs, teaching innovations, funding sources, researcher's interests, career opportunities, journals, and organizations useful to those in the field.[18] More than 20 formal graduate programs in environment and behavior are described.[19] Several, such as the Environmental Psychology Program at the City University of New York, and the School of Architecture and Urban Planning at the University of Wisconsin-Milwaukee, have extensive lists of publications based on strong ongoing research programs. Strong publication programs at universities are matched by a growing number of publishers developing a series of environment-behavior-design books.[20]

Another overview of the institutional developments is provided by Szigeti who describes the increasing activity exemplified by the new associations, sections, divisions, task forces, and journals which have appeared since the 60s.[21] Early and important is the Environmental Design Research Association (EDRA) formed in 1969 which will hold its 15th annual meeting in 1984. Each year EDRA has published the proceedings of these meetings and has over the years encouraged the development of networks of people interested in research and action on such topics as aging and the environment, childhood city, education in environment design research, energy and behavior, environmental/community psychology, environmental perception/cognition, environmental sociology, handicapped and the environment, interiors, pednet, post-occupancy evaluation, transnet,

---

18. Willo P. White, ed., *Resources in Environment and Behavior* (Washington, D.C.: American Psychological Association, 1979).

19. Gary W. Evans, "Graduate Programs," in *Resources in Environment and Behavior,* ed. W. P. White, pp. 39-92.

20. This includes Plenum with their series edited by Altman and Wohlwill, starting in 1976; Brooks/Cole and their series edited by Altman, Stokols and Wrightsman; and Hutchinson and Ross which has published some of the proceedings of the Environmental Design Research Association conferences and many other major volumes on environment-behavior-design topics.

21. Françoise Szigeti, "Environmental Psychology/Environment-Behavior: Overview of the Institutional Developments Since the 60s," *Design Research News* 12 no. 5 (1981), pp. 1-4.

and rural ecology. A more recent development is the formation of the International Association for People and Their Surroundings (IAPS) which had its first formal meeting in July 1982 at Barcelona, Spain.

In doing our reviews for *Progress in Human Geography*, we could see the results of this institutionalization of the field. We received a flood of mail in response to our request for reprints, preprints and references even though we limited the request to one year's production. Even after some selection there remained close to 300 references in each of the first two reviews. Byerts and White list 55 journals which publish articles on some aspect of environment and behavior.[22] This is by no means an exhaustive list for in our 2 reviews to date we referred to articles in an additional 92 journals not included on their list. This is close to the total of 150 journals, which Szigeti indicated, The Environmental Analysis Group (TEAG) take to stay abreast of the field.[23] Careful canvassing of foreign language journals would reveal many more which publish environment-behavior-design articles. The limits to the material covered in our reviews was dictated by the time available. At no time did we feel we had reached the limits of the published material.

Internationalization of Environment-Behavior-Design Research

From the signs of vitality of environmental perception research in North American geography the paper has broadened out to consider the institutionalization of environment-behavior-design research in a wide range of disciplines, most notably psychology and various design professions. One additional sign of this research area's vitality should be at least briefly noted, its international growth and spread.

The emergence of the International Association for People and Their Surroundings is an indication of European interest in environment-behavior-design research. The extent of this interest may also be gauged by consideration of a number of reviews of the French-speaking world,[24] Germany,[25] Italy,[26] and Spain.[27] There have

---

22. Thomas O. Byerts and Willo P. White, "Journals," in *Resources in Environment and Behavior*, ed. W. P. White, pp. 259-295.

23. See footnote 21.

24. Paul Claval, "L'évolution recente des recherches sur la perception," *Revista Geografica Italiana* 87 (1980): 6-24; Antoine S. Bailly, "Behavioral Geography in the French-Speaking World: Developments, Approach and Planning Experience," paper presented at the Institute of British Geographers conference in Leicester, January 5-8, 1981; see also Antoine S. Bailly and Bryn Greer-Wootten, "Behavioral Geography in Francophone Countries," *Progress in Human Geography* 7 (1983): 344-356.

25. Robert Geipel, "La géographie de la perception en Allemagne Fédérale," *Revista Geografica Italiana* 87 (1980): 88-95.

been special issues of Japanese journals or books devoted to environmental perception research[28] and work is underway on environmental psychology in the U.S.S.R.[29] A parallel and related development is the publication of the seven volume *Handbook of Cross-Cultural Psychology*.[30]

The extent of this international development was the topic of our third and final review of environmental perception research for *Progress in Human Geography*.[31] For this task we assembled within the course of a year a list of some 500 persons from areas other than English-speaking North America and Great Britain. The list contains contingents of 10 or more researchers from Japan, Germany, Switzerland, France, the Netherlands, U.S.S.R., Australia, Israel, Belgium, Sweden, South Africa, Finland, Egypt, Mexico and Italy. Included are geographers, psychologists, architects, sociologists, and planners and public agency officials from the full variety of professions usually associated with environment-behavior-design research in North America.

Some Comments on the Quality of the Research and Conclusions

By now I feel I have established that the level of research activity in environmental perception and in the entire environment-behavior-design field is higher than it has ever been. It is a research area of great diversity within geography where its practitioners have emerged from a great variety of origins, it is diverse in terms of the numbers of disciplines actively participating, and it is increasingly diverse in terms of international representation. The institutionalization of the field has provided a framework for a continuation of the momentum already generated, and the persistence of old environmental problems and emergence of new ones provides the need for continued efforts to improve the adjustments of people to their environments.

More important than any quantitative measures of growth in the

---

26. Felice Perussia, "La Percezione Dell'Ambiente: Una Ressegna Psicologica," in *Ricerca Geografica e Percezione Dell'Ambiente*, ed. Geipel et al. (Milano: Unicopli, 1980), pp. 51-67.

27. Jose E. Alvarez, "Gli Studi Sulla Percezione Ambiente Nella Geografia Spannola," *Revista Geografica Italiana* 87 (1980): 96-105.

28. [Editor unknown], "Mental Maps and Space Perception," *Chiri* (Geography) 25 (1980), entire issue; K. Nakamura, "Spatial Order and Its Cognition," *Notes on Theoretical Geography* 78 (March 1979), entire issue; M. Senda, "Territorial Possession in Ancient Japan: The Real and the Perceived," in *Geography of Japan*, ed. Association of Japanese Geographers (Tokyo: Teikoku-Shoin, 1980).

29. Toomas Niit, Juri Kruusvall, and Mati Heidmets, "Environmental Psychology in the Soviet Union," *Journal of Environmental Psychology* 1 (1981): 157-177.

30. Harry C. Triandis, ed., *Handbook of Cross-Cultural Psychology* (Boston: Alynn and Bacon, 1980), 7 volumes.

31. Saarinen, Sell, and Husband, see footnote 1.

field are some qualitative indicators of change and development. The broad acceptance within the geographic profession has already been noted. In addition, other social science professions have showed signs of acceptance of the environment-behavior-design field. Thus, for example, there have for several years been reviews of environmental psychology in the *Annual Review of Psychology*[32] and recently the environmental sociology group were given similar recognition in the *Annual Review of Sociology*.[33]

Other qualitative indicators of change and development which are noticed and documented in our reviews of the literature were an increasing integration of disciplines, and a more critical assessment of theories and methodologies than was evident a decade or more ago. Individuals from the separate disciplines are much more aware of corresponding fields in other environment-behavior-design disciplines and there were many papers which pull together the pertinent works for various combinations of subfields. As a result there are fewer isolated and parochial subgroups unaware of corresponding developments in the broader field. The importance of a strong base in empirical research is indicated by the appearance of many recent articles which have provided a critical assessment of specific methods, concepts, or assumptions, and even direct comparisons of several methods. Such articles need a substantial body of previous research to make direct comparisons possible. The papers falling into both of the categories considered in this paragraph tend to be written by seasoned researchers, active in the field for a decade or more.

A fine example of the cumulative weight of the trends, toward the increasing integration of disciplines and the more critical assessment of theories and methodologies, may be seen in the current effort to develop an agenda for environmental design research for the 1980s.[34] A task force assembled by the Environmental Design Research Association has representatives from specialty groups or divisions within the Human Factors Society, the American Planning Association of Applied Psychology, the Architectural Schools Research Centers Consortium, the American Society of Interior Designers, the Gerontological Society, the Environmental Design Research Association, the Association for the Study of Man-Environment Relations, the American Sociological Association,

---

32. Kenneth H. Craik, "Environmental Psychology," *Annual Review of Psychology* 24 (1973): 403-422; Daniel Stokols, "Environmental Psychology," *Annual Review of Psychology* 29 (1978): 253; James A. Russell and Lawrence M. Ward, "Environmental Psychology," *Annual Review of Psychology* 33 (1982): 651-688.

33. Riley E. Dunlap and William R. Catton, "Environmental Sociology," *Annual Review of Sociology* 5 (1979): 243-273.

34. Gary T. Moore, D. Paul Tuttle, Sandra C. Howell, and Donna Duerk, "Environmental Design Research for the 1980s: Toward an Agenda for Research," paper presented at the Environmental Design Research Association 12th Annual Conference, Ames, Iowa, April 2-6, 1981.

the Association of American Geographers, the American Psychological Association, the International Association for People and Their Surroundings, and the American Society of Landscape Architects. The project defines environmental design research to include the interactions between the human, social, built, and natural environments, and applications to improve the quality of the physical environment through public policy, planning, design, and education. The report to be produced represents a thrust toward further institutionalization of the field by defining goals and setting national priorities within their area of interest.[35]

Such a report would not have been possible a decade ago because of the incipient nature of the research. Today it seems a logical next step, likely to produce useful results. The difference is that environment-behavior-design research is now at the point where it can benefit not only from the vitality represented by the growing number and diversity of researchers, but also from the energy expended in the past. This intellectual capital is represented by the body of empirical research already generated, and by the increase in experience of many individual researchers who have developed with the field.

---

[35]. Gary T. Moore and Sandra C. Howell, *Environmental Design Research Directions for the 1980s: A Report from the Environmental Design Research Association,* second draft (November 1981).

CHAPTER 2
BEHAVIORAL GEOGRAPHY IN WESTERN EUROPE REFLECTIONS ON RESEARCH IN GREAT BRITAIN AND THE FRANCOPHONE NATIONS

JOHN R. GOLD

Over the last two years, I have been studying a cross-section of the literature produced by behavioural geographers in non-Anglophone Western Europe. I did so partly to redress the Anglo-American bias of my earlier reviews;[1] a bias already called into question by the publication in French, German, and Italian literature of a series of papers that collectively indicate a large and significant body of research which is virtually unknown by Anglophone geographers.[2] At the same time, I confess to having another, more parochial reason for undertaking this survey. For reasons to be discussed more fully later, the fortunes of behavioural geography are currently at a low ebb in Great Britain. Interest in behaviouralist research has declined markedly amongst geographers, with a serious diminution in the quality and quantity of research output. Quite apart from its intrinsic merit, therefore, the work of other European geographers clearly warranted the closest scrutiny as a possible source of new ideas and inspiration.

The contents of this paper reflect these aims. The first section briefly outlines the nature and scope of research by Francophone behavioural geographers--work that represents the largest single corpus of material from non-Anglophone Western

---

1. John R. Gold, "Teaching Behavioural Geography," *Journal of Higher Education* 1 (Spring 1977): 37-46; John R. Gold, *An Introduction to Behavioural Geography* (London: Oxford University Press, 1980).

2. Inter alia, see A.S. Bailly, *La perception de l'espace urbain* (Paris: Centre de Recherche de l'Urbanisme, 1977); R.E. Lob and H.W. Wehling, eds., *Geographie und Umwelt* (Kronberg: Verlag Anton Hain Meisenheim, 1977); R. Geipel, "La géographie de la perception en Allemagne Fédérale," *L'Espace Géographique* 7 (1978): 195-8; various authors, "Geografia della percezione," special issue, *Rivista Geografica Italiana* 87 (March 1980); A.S. Bailly, ed., *Percevoir l'espace: vers une géographie de l'espace vécu* (Geneva: Department of Geography, University of Geneva). For surveys of available sources, see my bibliographies: *Behavioural Geography in French Language Literature,* Public Administration series 1036, (Monticello, Ill.: Vance Bibliographies, 1982); and *Behavioural Geography in German and Italian-Language Literature,* Public Administration series, (Monticello, Ill.: Vance Bibliographies, forthcoming).

Europe. The second part seeks to place this research into a wider context by contrasting its progress to date with the experience of British behavioural geography at a comparable stage of development.

Behavioural Geography in Francophone Europe

The debt owed to the French School for upholding and maintaining humanistic traditions in geography is widely recognised, but the tendency has been for their contribution to be regarded as largely of historical significance. This opinion, however, has been challenged by two recent developments. First, a number of commentators[3] have pointed out the continuing relevance of French traditions for modern behavioural and humanistic geography. Secondly, there have been changes in French human geography itself. As late as the mid-1970s, for example, it was hard to demur from Claval's comment that "perception geography is of little importance in France" and that to find relevant research one needed to look to psychology, anthropology and sociology sources.[4] Yet a mere five years later, the same author was to record that perception studies had come to occupy a considerable place in contemporary geography.[5]

When examining the outpouring of research from geographers in Belgium, Quebec, France and Switzerland that had prompted this reappraisal, one finds a large volume but lack of programmatic direction. Lack of space means that it is impossible to present a comprehensive survey here, but general remarks may be made about the outlook, methodology, and empirical content of Francophone studies.

In common with behavioural geography elsewhere in the world, Francophone research has developed a strong multidisciplinary outlook. To date the most productive links have been with psychology. Besides having the heritage of such well-known writings as those of Piaget and Merleau-Ponty, Francophone geographers have benefited from close contacts with the emerging field of "environmental psychology" (which the French term the "psychology of space"). The strength of the cognitive approach within French psychology, the emphasis laid upon the role of information in behavioural processes, and the attention given to the temporal

---

3. Anne Buttimer, *Society and Milieu in the French Geographical Tradition* (Chicago: Rand McNally, 1971); Vincent Berdoulay, "French Possibilism as a Form of Neo-Kantian Philosophy," *Proceedings of the Association of American Geographers* 8 (1976): 176-9; David Ley and Marwyn S. Samuels, eds. *Humanistic Geography: Prospects and Problems* (London: Croom Helm, 1978); Derek Gregory, "Human Agency and Human Geography," *Transactions of the Institute of British Geographers* (New Series) 6 (March 1981): 1-18.

4. Paul Claval, "Contemporary Human Geography in France," *Progress in Geography* 7 (1975): 268.

5. Paul Claval, "L'évolution recente des recherches sur la perception," *Rivista Geografica Italiana* 87 (March 1980): 21.

dimension of experience and behaviour[6] have all been valuable influences upon the emerging shape of Francophone behavioural geography, and there have been numerous instances on which the interests of geographers and psychologists have converged.[7] Where Francophone behavioural geography perhaps differs from similar studies elsewhere lies in the range of contacts that have been built up across the full spectrum of the social sciences, arts, and humanities. As Bailly[8] demonstrates, links have been developed with such disciplines as structural anthropology, ethnology, sociology, political theory, philosophy (especially radical humanism and Marxism, linguistics, media studies, and urban design (particularly semiological theory).

This breadth of outlook has contributed much to the methodological eclecticism of Francophone behavioural geography. Leanings toward the social and behavioural sciences have encouraged use of a variety of scaling and psychometric techniques, with considerable interest in multivariate analyses. Studies adopting such techniques[9] will seem reasonably familiar to Anglophone geographers, since heavy use is made of work already pioneered by Anglo-American researchers. Rather less familiar, however, may well be the range of methodologies that have been culled from the arts and humanities. Considerable interest, for example, has centered upon the potential of semiotic analyses to investigate environmental meaning,[10] even if tangible results are as yet meagre. Another important node of effort has been to employ methods culled from literary criticism, mass communication research and art history to extract environmental imagery from secondary sources--including novels and popular literature, letters and diaries, film and documentary art.[11]

---

6. A.S. Bailly, "La géographie de la perception dans le monde francophone," *Geographica Helvetica* 36 (March 1981): 15.

7. To gain some idea of the potential of environmental psychology in this respect, see J. Pailhous *La représentation de l'espace urbain: l'exemple du chauffeur de taxi* (Paris: Presses Universitaires de France, 1970); A. Moles and E. Rohmer, *Psychologie de l'espace* (Paris: Castermann, 1972); C. Levy-Leboyer, *Psychologie et environnement* (Paris: Presses Universitaires de France, 1980); G.N. Fischer, *La psychologie de l'espace* (Paris: Presses Universitaires de France, 1981).

8. A.S. Bailly, loc. cit., 15-17.

9. Ibid., 19-20.

10. Various authors, "Paysages et sémiologie," special isssue, *L'Espace Géographique* 2 (June 1974): 113-51; A.J. Greimas, *Sémiotique et sciences sociales*. (Paris: Le Seuil, 1976); P. Auchlin, "Ralentir enfants! Vers un approche sémiologique de l'analyse de l'espace à travers des dessins d'enfants," *Geographica Helvetica* 33 (1978): 67-74.

11. Inter alia, J. Seebacher, "L'espace vécu des personnages de Madame Bovary," *Cahiers de Géographie de Caen* supplement 1 (1973): 59-66; L. Dufour, "La météorologie dans les Mémoires de Saint-Simon," *Météorologie* 6 (1976): 183-188; J.P. Guerin and H. Gumuchian, "Les mythologies de la montagne: étude comparée de

With regard to subject matter, it goes almost without saying that, by virtue of dealing with geographical regions little studied by English-speaking geographers, Francophone research offers a rich corpus of new empirical data. But the issue is far more than just one of increasing the spatial coverage of research. In the process of looking at different areas, different questions are necessarily posed. For example, the main focus of Francophone natural hazards research has been human adjustment and response to drought in the Sahel, an arid and under-developed region of West Africa, characterised by a pastoral economy and inhabited principally by nomads.[12] Not surprisingly, the findings contrast strikingly with those of Anglo-American research, with its emphasis upon flood hazards and preoccupation with developed areas--particularly in the USA itself.[13] By the same token, Francophone research on the perceptions and behaviour of new urban migrants centres upon groups that rarely figure in Anglo-American work, such as North African and Arabic-speaking communities; people who frequently come from rural areas are semi-literate and have very different social structures than those of the host society. The data produced by the rapidly growing number of studies on this subject[14] can help to shed new light on the processes by which people come to terms with the urban environment.

Before concluding this brief survey, I wish to make a final point about the intellectual leanings of behavioural geography in Francophone Western Europe. Janus-like, it looks simultaneously in two directions; outwards towards research ideas appearing in the Anglophone world and inwards towards long-standing French

---

deux textes publicitaires (application de la methode AAAD 75)," *Revue de Géographie Alpine* 65 (1977): 385-402; R. Kleinschmager, "La ville et le cinéma: recherches sur la représentation de l'espace urbain dans le cinéma européen et nord-américain," *L'Espace Géographique* 7 (June 1978): 126-130; J. Wintsch, "Marguerite Yourcenar--'Mémoires d'Hadrien': Essai d'interprétation du rôle de l'espace dans un roman," *Brouillons Dupont* 4 (1979): 59-72; M. Chevalier, "Chronique bourguignonne: la vie régionale de la Bourgogne à travers une 'lettre d'information'," *Revue Géographique de L'est* 20 (1980): 229-237.

12. J. Gallais, "Une contribution a la connaissance de la perception spatiale chez les pasteurs du Sahel," *L'Espace Géographie* 5 (March 1976); M. Museur and R. Pirson, "Conséquences de la secheresse sur milieu saharien: l'Ahaggar," *Revue Belge de Géographie* 100 (1976): 293-311; various authors, "Le Sahel," *Travaux et Documents de Géographie Tropicale* 30, special issue (1977): 1-284; J. Gallais, "Histoire et thèmes d'une recherche au Sahel," *Cahiers Géographiques de Rouen* 15 (1981): 3-10.

13. A point emphasised by W.I. Torry, "Hazards, Hazes and Holes: A Critique of 'The Environment as Hazard' and General Reflections on Disaster Research," *Canadian Geographer* 23 (1979): 368.

14. V. Patte, "Parler en France: les possibilités des femmes immigrées," *Méditerranéan Peoples* 8 (1979): 78-88; X. Piolle, *Les citadins et leur ville: approche de phénomènes urbains et recherche méthodologique* (Toulouse: Privat, 1979); A. Schule, "Le centre de Tours et les Tourangeaux: réflexion sur l'espace vécu," *Cahiers de la Loire Moyenne* 10-11 (1979-1980): 69-85; E.S. Santana, "Les immigrés dans la ville: analyse d'un espace écologique à Toulouse," *Revue Géographique des Pyrénées et du Sud-Ouest* 51 (1980): 137-151.

geographical traditions. Studies of urban perception and behaviour, which constitute one of the two major foci of research, does rest heavily upon Anglo-American work, (a source of influence partly mediated by the writings of French Canadian geographers, with their ready access to both French and English language literature). By contrast, work at the regional scale--the other main core--is much more an extension of the traditional focus of French human geography and can only be understood fully within that tradition. For example, a concept often used in this research is that of *l'espace vécu*.[15] As Sarre[16] has noted, *l'espace vécu* has no direct English equivalent, involving as it does, an amalgam of perceived space and life space, and asserting the intimate links between people and place. Its significance as a framework for the study of regions of various types and sizes, rural and urban, needs to be seen in the light of broader intellectual currents pertaining in French geography. Although this may serve to make this aspect of research somewhat impenetrable to non-Francophone geographers, it undoubtedly does serve to supply a continuity with the past which is not so readily apparent in Anglo-American behavioural geography.

Through a Glass, Darkly

This last point may be developed further. When looking at Francophone research from the standpoint of the outsider looking in, one is struck by the parallels that can be drawn with earlier British experience. Naturally there are contrasts, and this question of continuity with the past provides an apposite example. In Britain we stressed the discontinuous, even revolutionary, nature of our work compared with what went before and, in doing so, perhaps sacrificed a source of strength that has been retained by our counter parts in Francophone Europe. Nevertheless, I cannot help but discern certain similarities between the progress of their research and the situation to be found in Britain some ten years ago. Looking back to that time, it is possible to identify close similarities with what one regards as the most promising features of present-day Francophone research: the same dynamism and outward-looking attitude, a desire to collaborate with other disciplines, a belief that today's exploratory studies would lead to

---

15. A. Frémont, *La région: essai sur l'espace vécu*, (Saint-Brieuc: Presses Universitaires de Bretagne 1972); A. Frémont, *La région, espace vécu* (Presses Universitaires de France, 1976); J. Gallais, "Des quelques aspects de l'espace vécu dans les civilisations du monde tropical," *L'Espace Géographie* 5 (March 1976): 5-10; J.L. Piveteau, "L'espace vécu chez le peuple Hébreu," *Geographica Helvetica* 33 (1978): 141-144.

16. P.V. Sarre, "Contrasts and Similarities between Anglophone and Francophone Approaches to Environmental Perception and Behavioural Geography," (mimeo.), Milton Keynes: Department of Geography, Open University, 1982, 1.

tomorrow's conceptual and methodological advances and, above all, the development of a consolidated body of geographical knowledge.

In the case of Great Britain, these were to prove pious hopes. It is almost melancholy to note that the highwater mark for British behavioural geography probably came in 1973 with the invited seminar held at Brunel University under the auspices of the Social Science Research Council. At that meeting, the audience of geographers and psychologists held a series of highly productive sessions at which they outlined progress in their respective fields, compared notes on common themes and trends, identified research priorities, and generally declared their interests in working together. Further plans were noted for subsequent meetings and as the seminar's final report so succinctly put it:[17]

> the number of follow-up plans which have been or are being put underway is a measure of the success of the seminar in furthering environmental perception as an active research field in this country.

Fine sentiments, but in practice there was little net result. Consciously or otherwise, the environmental psychologists decided to go their own way, steadily building their subject into a formal sub-discipline and pursuing contract research. Meanwhile, the geographers, who had hitherto been content to come together under the banner of 'environmental perception', began to split into schismatic groups. The available research lost its cutting edge and, of late, one might suppose that the only matter of any real concern was the regional consciousness of elite novelists and literati. Ritual denunciations of behaviouralist research in the geographical press, some verging on caricature, went unanswered--a fair indicator of the general state of health of this field of inquiry.

In searching for the reasons why British behavioural geography is currently in the doldrums, it is conventional to think of problems directly related to the research itself, such as unrealistic initial expectations, tensions between humanist and positivist orientations, lack of attention to the development of methodology and inadequate channels for research publication. Yet important as these factors have been, I do feel that there is another contributing factor that has been largely overlooked, namely, the relative isolation of British behavioural geographers. For complex institutional reasons linked, in large measure, with the timing of the expansion of higher education and of the consequent

---

17. Steering Committee, Report to the Human Geography and Psychology Committees of the Social Science Research Council on the Seminar on Environmental Perception held at Brunel University, 4-6 May 1973 (mimeo.), London: Social Science Research Council, 1973.

recruitment of staff for geography departments, British human geography remains firmly rooted in mainstream spatial analysis. Whereas it is not uncommon to find the occasional individual who lists 'behavioural geography' as his or her research interest, the bulk of British work in this field stemmed from the staff and graduate students of a small number of academic institutions, such as the Geography Departments of the Universities of Durham and Bristol, the Centre for Urban and Regional Studies at the University of Birmingham, and the Department of Geography of University College, London. When these institutions lost interest in this area of study in the mid-1970s or turned to other preoccupations when key staff departed, there was a distinct slackening in the pace of development of British behavioural geography.

Returning to a comparative perspective, it is here that nagging doubts arise. Research in Francophone Europe currently emanates from individual researchers based at a small number of widely scattered universities, notably Lausanne and Geneva in Switzerland, Liege in Belgium, and Montpellier, Rouen, Grenoble, Caen and Strasbourg in France. Although the researchers themselves are prolific in their output, they tend to work in isolation from one another and, with the added problem of linguistic barriers, from colleagues elsewhere in Europe. The British experience showed that a similarly small and isolated group of researchers were unable to maintain the progress that earlier research suggested might be forthcoming. Certainly they were never able to build up that infrastructure of newsletters, periodicals, conferences and symposia which are so often important in the development of a field of study.[18] Whether the Francophone geographers can fare better than their British counterparts remains to be seen, but one suspects that their work is now approaching this critical point.

Acknowledgements

This is a slightly revised version of a paper first given at the 78th Annual Meeting of the Association of American Geographers at San Antonio, Texas. I am grateful to the Overseas Conference Fund of the British Academy which made my attendance at this meeting possible. I also wish to thank Drs. Philip Sarre and Antoine Bailly for supplying me with invaluable research materials.

---

18. In making this point, I am extending the comments made by T.R. Lee, *Psychology and the Environment* (London: Methuen, 1976) p. 16.

CHAPTER 3
HAZARD PERCEPTION STUDIES:
CONVERGENT CONCERNS AND DIVERGENT APPROACHES DURING THE PAST DECADE

JAMES K. MITCHELL

Perception is a central, but nebulous and controversial, concept in Hazard Research. Its value as an organizing principle has been variously misunderstood,[1] questioned,[2] and assailed[3] by academic observers of the field. Although natural scientists are increasingly receptive to perception as a bridging and unifying concept for environmental management research, the same cannot be said for public policy makers and personnel in hazard management agencies.[4] Insofar as its use is concerned, the term "perception" has found little favor outside of research reports and books by academics and private consultants. Nonetheless, a significant number of geographers, psychologists, sociologists and others engaged in the interpretation of human response to hazard subscribe to models of adjustment choice and risk mitigation wherein perception of hazard risks and perception of adjustments to hazard are fundamental components.[5] (Figs. 3.1 and 3.2)

---

1. Reginald C. Golledge, "Misconceptions, Misinterpretations, and Misperceptions of Behavioral Approaches in Human Geography," *Environment and Planning,* series A 13 (November 1981): 1325-1344.

2. Timothy O'Riordan, *Environmentalism* (London: Pion Press, 1976).

3. Trudi Bunting and Leonard Guelke, "Behavioral and Perception Geography: A Critical Appraisal," *Annals of the Association of American Geographers* 69 (September 1979): 448-462. See also: Thomas F. Saarinen, "Commentary--Critique of Bunting-Guelke Paper," *Annals of the Association of American Geographers* 69 (September 1979): 465-468; and Trudi Bunting and Leonard Guelke, "Commentary in Reply," *Annals of the Association of American Geographers* 69 (September 1979): 471-474.

4. M.J. Clark, "Geomorphology in Coastal Zone Management," *Geography* 63 (1978): 273-282; Dennis J. Parker and Donald M. Harding, "Natural Hazard Evaluation, Perception and Adjustment," *Geography* 64 (1979): 307-316; Robert H. Simpson and Herbert Riehl, *The Hurricane and Its Impact* (Baton Rouge: Louisiana State University Press, 1981), pp. 288-296.

5. Robert W. Kates, *Natural Hazard in Human Ecological Perspective: Hypothesis and Models,* Natural Hazard Research Working Paper no. 14 (Worcester: Clark University, Department of Geography, 1970); reprinted in *Economic Geography* (July 1971); Dennis S. Mileti, "Human Adjustment to the Risk of Environmental Extremes," *Sociology and Social Research* 65 (April 1980): 327-347.

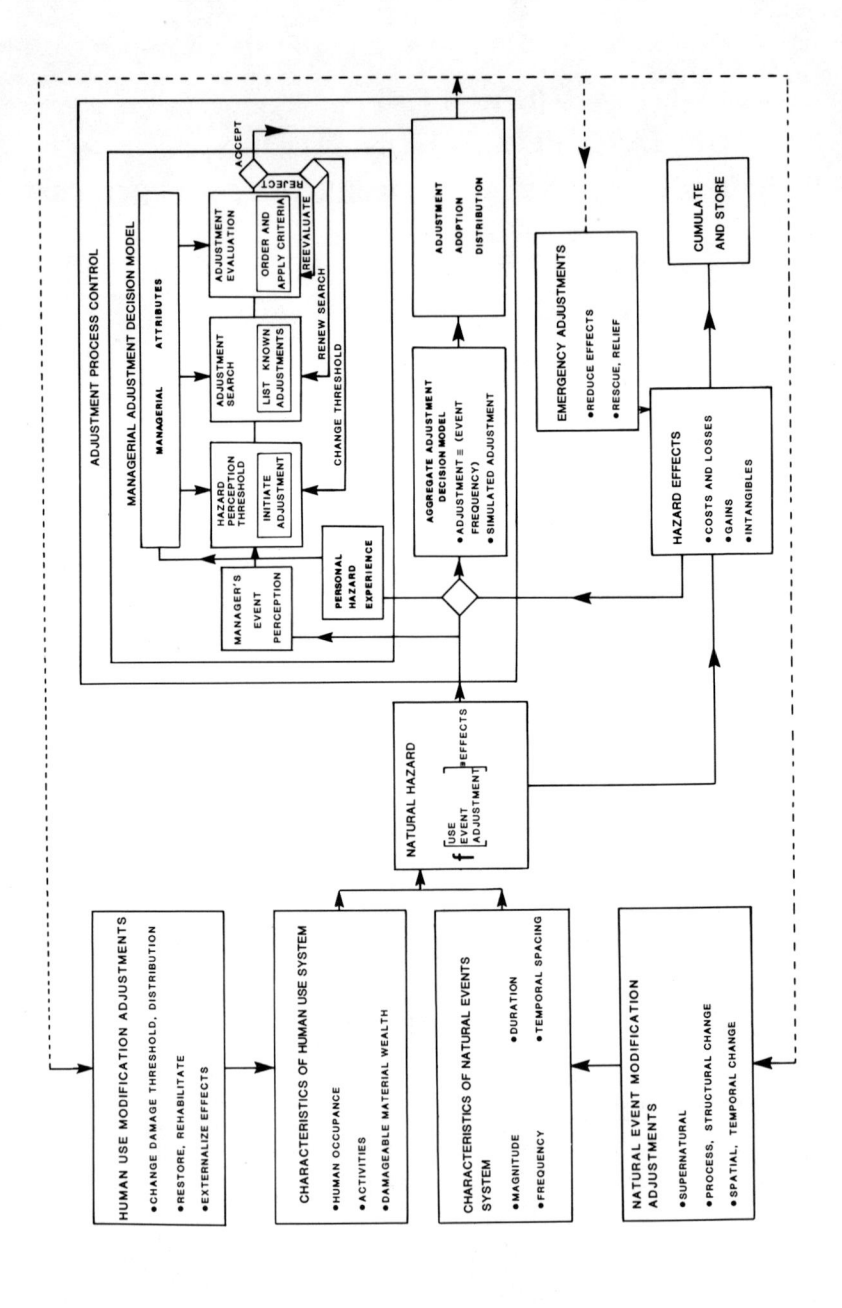

Figure 3.1 Human Adjustments to Natural Hazards (After Kates, 1970)

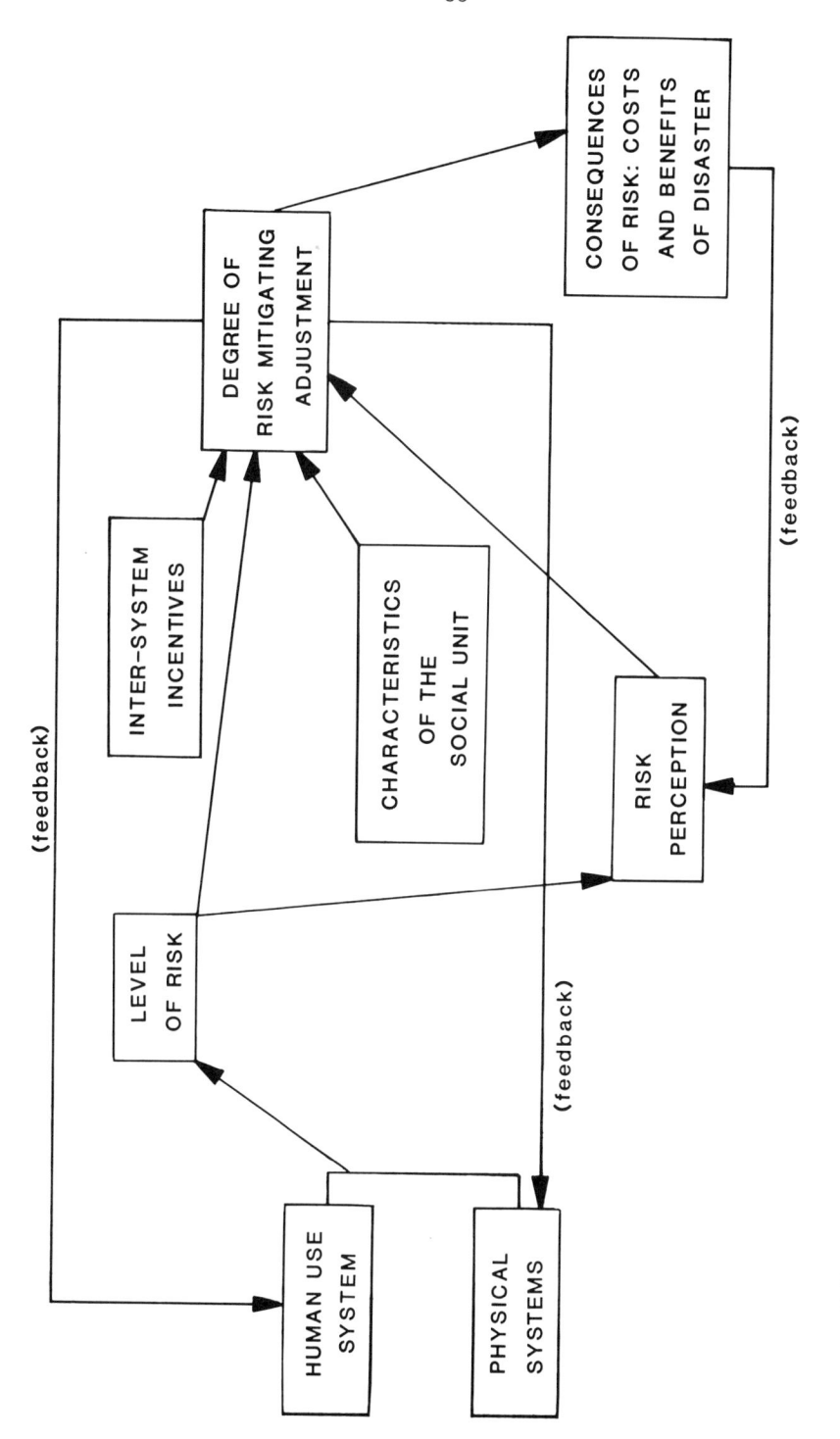

Figure 3.2 Key Concepts in the Emerging Theory of Risk Mitigating Adjustment (After Mileti, 1980)

During the past decade the volume of published research on Hazard Perception has been relatively modest in comparison to the total amount of work in Hazard Research. To some extent this reflects changes in the general field of Hazard Research which have increased the amount of attention paid to: (1) collective influences on hazard decisions; and (2) constraints which limit the hazard manager's freedom of action to adopt appropriate adjustments. Most studies can be classified as: (1) assessments of hazard awareness, or (2) explanations of perceptual processes which highlight the roles of personality or sources of judgemental bias. Emphasis has been placed on awareness and responses of indigenous groups of domestic or local hazards rather than on problems of international or foreign hazard management.

What is Hazard Perception

One of the first tasks of a reviewer is to define the subject under study. Unfortunately, there is no generally agreed on definition of Hazard Perception. The word "perception" is found in a great many published hazard studies but with little consistency of usage. In some it refers to risk assessments made by potential victims.[6] In other concerns: attitudes to environment;[7] reported information about hazards;[8] awareness of physical processes contributing to hazard;[9] "comprehension of the character of hostile environments;"[10] and, identification of adjustments to hazard.[11] The "perceptions" of victims, scientific experts, public officials, hazard interest group members, and many other people are variously subject to scrutiny. Terms like "hazard cognition" or "hazard recognition" are also used to describe phenomena and processes that are elsewhere labelled "hazard perception."[12]

---

6. M. Greene, R. Perry, and M. Lindell, "The March 1980 Eruptions of Mt. St. Helens: Citizen Perceptions of Volcano Threat," *Disasters* 5 (1981): 49-66; Glenn Shiffee, Jeffrey Burroughs, and Stuart Wakefield, "Dissonance Theory Revisited: Perception of Environmental Hazards in Residential Areas,"·*Environment and Behavior* 12 (March 1980): 33-51.

7. S.P. Malhotra, "Traditional Perceptabilities of Environment and Desertification: A Case Study," *Economic Geography* 53 (October 1979): 347-352.

8. R.D. Needham and J.G. Nelson, "Newspaper Response to Flood and Erosion Hazards on the North Lake Erie Shore," *Environmental Management* 1 (1977): 521-540; K.F. Gregory and R.F. Williams, "Physical Geography from the Newspaper," *Geography* 66 (January 1981): 42-52.

9. D.A. Eastwood and R.W.G. Carter, "The Irish Dune Consumer," *Journal of Leisure Research* 13 (1981): 273-281.

10. Sen-dou Chang, "Chinese Perception and Transformation of Arid Lands," *The Chinese Geographer* 10 (Spring 1978): 1-12.

11. Edgar L. Jackson, "Response to Earthquake Hazard: The West Coast of North America," *Environment and Behavior* 13 (July 1981): 387-416.

12. Perry O. Hanson, John D. Vitek, and Susan Hanson, "Awareness of Tornadoes: The Importance of an Historic Event," *Journal of Geography* 78 (January 1979): 22-25; Milton E. Harvey, John W. Frazier, and Mindaugas Matulionis, "Cognition of a Hazardous Environment: Reactions to Buffalo Airport Noise," *Economic Geography* 55 (October 1979): 263-286.

Formal definitions of the field are also at variance. Some include natural events and human responses. White's working definition of hazard perception is ". . . the individual organization of stimuli relating to an extreme event or a human adjustment," but he cautions that ". . . awareness and judgment of natural events and human opportunities encompass[es] far more" . . . than that statement implies.[13] Mileti, Drabek and Haas offer a different formulation based on their assessment of the terms' use in published literature. This definition omits mention of adjustments: "Perception of hazard is an individual's understanding of the character and relevance of a hazard for self and/or community."[14] Mileti also defines risk perception as ". . . cognition or belief in the seriousness of the threat of an environmental extreme, as well as the subjective probability of experiencing a damaging environmental extreme."[15] This implies that a second definition, relating to perception of adjustments, is a necessary component of a comprehensive hazard perception definition. These definitions share a concern for individual awareness of extreme events. Yet one also stresses the prospects of individual losses, another focuses on community consequences, and the third includes perceived adjustments as well as perceived threats. Other scholars have sought to bring "collective and cultural perceptions" within the boundaries of the field.[16] Clearly, the concept of hazard perception has different meanings for different people. This uncertainty is reflected in both the breadth of the hazard perception literature and the exploratory nature of studies in many areas.

Hazard Perception Studies are a branch of the much larger field of Hazard Research. The latter is concerned with the totality of factors which generate, sustain, exacerbate, or mitigate those characteristics of natural and man-made environments that threaten human safety, emotional security, and material well-being. Although students of Hazard Perception often seek guidance from--and relate findings to--Environmental Perception Research, their primary reference group is the Hazard Research community. Moreover, the salience of perception studies within Hazard Research has changed

---

13. Gilbert F. White, ed., *Natural Hazards: Local, National, Global* (New York: Oxford University Press, 1974), p. 4.

14. Dennis S. Mileti, Thomas E. Drabek, and J. Eugene Haas, *Human Systems in Extreme Environments: A Sociological Perspective,* Program on Technology, Environment and Man Monograph no. 21 (Boulder: University of Colorado, Institute of Behavioral Science, 1975), p. 23.

15. Mileti, loc. cit., p. 336.

16. Parker and Harding, loc. cit., p. 307.

markedly over the decades since White's early work on flooding.[17] Hazard perception studies were essentially unknown until Kates' pathbreaking analysis of *Hazard and Choice Perception in Flood Plain Management*.[18] Thereafter, perception studies grew to dominate the parent field in the late 1960s and early 1970s. More recently, their relative significance has somewhat declined as hazards researchers explore alternative theoretical and methodological perspectives derived from a variety of newly cooperating disciplines. Nevertheless, Perception Studies are still an important, and developing, element of Hazard Research. In view of these changing relationships an examination of recent developments in Hazard Research is in order.

Recent Developments in Hazard Research (1972-1982)

By 1972 Hazard Research was already an interdisciplinary field. Therein, geographers, psychologists, economists, engineers, insurance specialists, and community planners directed attention to complex policy-related problems involving a wide range of extreme geological and atmospheric hazards in a variety of nations. Hazard typologies, classifications and models had been formulated and the groundwork had been laid for a hazard adjustment theory based on concepts of boundedly rational decision making under conditions of uncertainty. National and international scientific bodies were providing research sponsorship and financial support. By the time members of the International Geographical Union's Commission on Man and Environment assembled in Calgary (1972) to report on the results of collaborative cross-national and cross-cultural hazard adjustment surveys, the stage was set for accelerated expansion of activities.[19]

Looking back across the succeeding decade it is clear that the promise of major transformations was indeed realized. Changes occurred on many different levels. What had been Natural Hazard Research gradually became Hazard Research as technological problems like nuclear, transportation and workplace accidents; toxic substances; explosions; airport noise; and global atmospheric

---

17. Gilbert F. White, *Human Adjustment of Floods: A Geographical Approach to the Flood Problem in the United States,* University of Chicago, Department of Geography Research Papers, no. 29 (Chicago: University of Chicago, Department of Geography, 1945).

18. Robert W. Kates, *Hazard and Choice Perception in Flood Plain Management,* University of Chicago, Department of Geography Research Papers, no. 78 (Chicago: University of Chicago, Department of Geography, 1962).

19. J.K. Mitchell, "Natural Hazards Research," in *Perspectives on Environment,* ed. Ian R. Manners and Marvin W. Mikesell (Washington, D.C.: Association of American Geographers, 1974), pp. 311-341.

pollution were added to the list of subjects under investigation.20 Initial work on social hazards such as violence and mental illness also began to appear.21

Paralleling this transition was a marked increase in the amount of attention paid to users of hazard research results. Hazard Research began as an academic based endeavor aimed at influencing federal resource management agencies and other implementors of national policy. Throughout the 1970s it grew to include specialized training sessions, publications and conferences variously directed towards legislators, civil defense personnel, specific research communities like the National Sea Grant Program, the communications media, disaster victims, public interest groups, bankers, insurance experts, real estate developers, planners, private consultants, state and local officials, and many other user groups.

As the number of active researchers grew, the pace and range of Hazard Research also expanded in other ways. This was especially noticeable among geographers but other disciplines recorded similar, though less accentuated, trends. For example, annual meetings of the Association of American Geographers routinely included multiple, well attended sessions devoted to hazard topics. Geographers authored and edited half a dozen influential books which reported major research results and charted future lines of advance.22 Much of the work was translated into college level texts to serve a

---

20. Christoph Hohenemser, Roger Kasperson, and Robert Kates, "The Distrust of Nuclear Power," *Science* 196 (1 April 1977): 25-34; Thomas Bick and Christoph Hohenemser, "Target: Highway Risks II--The Government Regulators," *Environment* 21 no. 2, (March, 1979), pp. 6-15 and 29-38; Baruch Fischoff, Christoph Hohenemser, Roger E. Kasperson, and Robert W. Kates, "Handling Hazards," *Environment* 20 (September 1978): 16-20 and 32-37; Robert C. Harriss and Christoph Hohenemser, "Mercury: Measuring and Managing the Risk," *Environment* 20 (November 1978): 25-36; Donald Zeigler, Stanley D. Brunn, and James H. Johnson, Jr., "Evacuation from a Nuclear Technological Disaster," *Geographical Review* 71 (January 1981): 1-16; Kent Barnes, James Brosius, Susan C. Cutter, and James K. Mitchell, *Responses of Impacted Populations to the Three Mile Island Nuclear Accident: An Initial Assessment*, Rutgers University Department of Geography Discussion Papers, no. 13 (New Brunswick: Rutgers University, Department of Geography, 1979); Susan Cutter, Stephen Decter, James Brosius, and Charles Kelly, *Institutional and Individual Responses to Toxic Chemical Fires: The Chemical Control Corporation Fire, April 21, 1980, Elizabeth, New Jersey*, Rutgers University Department of Geography Discussion Papers, no. 17 (New Brunswick: Rutgers University Department of Geography, 1980).

21. James K. Mitchell, "Social Violence in Northern Ireland," *Geographical Review* 69 (April 1979): 179-201; James K. Mitchell, "Adjustment to Social Violence: The Northern Ireland Tourist Industry," unpublished paper delivered at the Annual Meeting of the Association of American Geographers Philadelphia, 1979; James K. Mitchell, "Getting Away From It All: Impacts of Social Violence on Tourism in Northern Ireland," Rutgers University Department of Geography Discussion Paper (New Brunswick, Rutgers University, Department of Geography, 1982 forthcoming); Julian Wolpert, "The Dignity of Risk," *Transactions of the Institute of British Geographers* 5 (1980): 391-401.

22. White, *Natural Hazards*; Gilbert F. White and J. Eugene Haas, *Assessment of Research on Natural Hazards* (Cambridge: MIT Press, 1975); J.E. Haas, L. Bowden, and R.W. Kates, *Reconstruction Following Disaster* (Cambridge: MIT Press, 1977); Ian Burton, Robert W. Kates, and Gilbert G. White, *The Environment as Hazard* (New York: Oxford University Press, 1978); R.L. Heathcote and B.G. Thom, eds., *Natural Hazards in Australia* (Canberra: Australian Academy of Sciences, 1979).

growing body of students enrolled in hazards courses.[23]

Geographers were also instrumental in establishing new hazards research institutions and founding specialized hazards publications outlets. At the University of Colorado, hazard research became a major activity, first within the Institute of Behavioral Science, and later in a government and university supported national clearinghouse for information pertaining to hazards (i.e., Natural Hazards Research and Applications Information Center, 1976). The Center conducts research and provides assistance to public and private hazards interest groups. It also organizes influential national Workshops on Natural Hazards which each year involve several hundred invited participants from academic, government agencies, business organizations and elsewhere. Since 1972, under the aegis of the Center and the Institute of Behavioral Science, twenty one (21) new titles have been added to the Natural Hazard Research Working Paper Series, together with thirty four (34) monographs, a hazard research bibliography series, a bimonthly newsletter *(Natural Hazards Observer)*, and other assorted publications.

At Clark University, geographers who had helped to develop the field of Natural Hazards Research turned their attention to technological hazards and formed a Hazard Assessment Group. In addition to geographers this included physicists, engineers, economists, philosophers and environmental scientists. A series of Working Papers was published dealing with comparative hazard sources, costs and management expenditures.[24] With the receipt of major support from the National Science Foundation and other sources, a Center for Technology, Environment and Development was created. Much of the Center's Hazard Research has reached a wide audience through the pages of the mass circulation journal *Environment*.[25] Geographers from the Clark group are represented on

---

23. John Whittow, *Disasters: The Anatomy of Environmental Hazards* (Athens: University of Georgia Press, 1979); Harold D. Foster, *Disaster Planning: The Preservation of Life and Property* (New York: Springer-Verlag, 1980); A.H. Perry, *Environmental Hazards in the British Isles* (London: George Allen and Unwin, 1981); K. Smith and Graham Tobin, *Human Adjustment to the Flood Hazard* (London: Longman, 1979).

24. J. Tuller, "The Scope of Hazard Management Expenditure in the U.S.," working paper, Hazard Assessment Group, Clark University, Worcester, 1978; R.L. Goble, M. Yersel, and C. Hohenemser, "Estimates of Direct Costs and Productivity Losses Due to Technological Hazard Mortality," working paper, Hazard Assessment Group, Clark University, Worcester, 1978.

25. Robert C. Harriss, Christoph Hohenemser, and Robert W. Kates, "Our Hazardous Environment," *Environment* 20 (November 1978): 25-36; Thomas Bick and Roger E. Kasperson, "Pitfalls of Hazard Management," *Environment* 20 (October 1978): 30-42; Paul Slovic, Baruch Fischhoff, and Sarah Lichtenstein, "Rating the Risks," *Environment* 21 (April 1978): 14-20, 36-39; Patrick Derr, Robert Goble, Roger E. Kasperson, and Robert W. Kates, "Worker/Public Protection: The Double Standard," *Environment* 23 (September 1981): 6-15; Julie Graham and Don Shakow, "Risk and Reward: Hazard Pay for Workers," *Environment* 23 (October 1981): 14-24; Dale R. Hattis, Robert Goble, and Nicholas Ashford, "Airborne Lead: A Clearcut Case of Differential Protection," *Environment* 24 (January/February 1982): 14-24; Mary Melville, "Risks on the Job: The Worker's Right to Know," *Environment* 23 (November 1981): 12-20, 42-45.

the editorial boards of *Environment* and the new journal, *Risk Analysis*. They have also been among the leaders of an international, multi-disciplinary effort to establish a Society for Risk Analysis.

A third approach to hazards research has developed at the University of Toronto's Institute for Environmental Studies. This focuses on environmental perception and emergency planning research. The Institute's Director is also Chairman of the International Geographical Union's Working Group on Perception of the Environment--a successor to the Commission on Man and Environment. With support from UNESCO's Man and the Biosphere Program and other organizations, the Group has published a series of Working Papers and sponsored a Workshop in Ibadan, Nigeria (1978). Six of the nine Working Papers, and fifteen (15) of the twenty eight (28) papers delivered at Ibadan dealt with hazards problems. The Working Group has decided to focus its future activities on interrelationships between Environmental Hazards and Development, the management of pests and pesticides, and the perception of highly valued landscapes. In a separate initiative, Emergency Planning Canada has provided support for an Emergency Planning Program at the Institute and a series of Working Papers on Emergency and Risk Research.

Geographers from the three institutions noted above have joined with other academics and government scientists to undertake additional hazard studies. For example, in 1973 the Special Committee on Problems of the Environment (ICSU) initiated a research project on the Communication of Environmental Information and Societal Assessment and Response. This quickly focused on the concepts of environmental risk and risk assessment. Two important reports have emerged from various project working group meetings.[26] These review procedures for identifying risks associated with environmental or technological hazards; risk assessment methods; international case studies of specific risk management problems; the formulation of public policies on risk and emerging problems.

Although the bulk of recent geographical work on hazards emanates from Colorado, Clark and Toronto, smaller groups of hazards geographers exist at such places as the University of Victoria, University of Waterloo, University of Wisconsin and Rutgers University. Individual scholars can also be found at many other institutions. Curiously, despite the pace and scale of this activity, Hazard Research has not received complete recognition

---

26. Robert W. Kates, *Risk Assessment of Environmental Hazard,* SCOPE Report no. 8 (New York: John Wiley and Sons, 1978); Anne V. Whyte and I. Burton, eds., *Environmental Risk Assessment,* SCOPE Report no. 15 (New York: John Wiley and Sons, 1980).

within the discipline of geography in North America. Hazard Research subjects are underrepresented in the pages of major disciplinary journals and at least one recent benchmark summary of major currents in the discipline completely failed to mention the field.[27]

The Hazard Research tide has swept well beyond the shores of North America to encompass scholars in Europe, Australia, New Zealand, Japan, India and a variety of other nations. Hazard geographers are particularly active in the United Kingdom and Australia. In 1973, together with architects and economists, geographers helped to set up the now defunct Disaster Research Unit at the University of Bradford. This organization published a series of Occasional Papers that followed somewhat different philosophical and methodological approaches to hazards than those of most North American researchers. The implications of these differences are discussed at greater length below. Although the Disaster Research Unit's principal researchers accepted positions with the University of Bath, Oxfam and Clark University, their initiative was continued by the International Disaster Institute in London. Since 1973 the Institute has issued eighteen (18) detailed reports on topics including: drought in Africa; earthquakes in Italy and Turkey; flooding in London; refugees in Pakistan, Cambodia, Thailand and Honduras; nuclear reactors; and disaster training institutes. These incorporate the contributions of specialists in public health, nutrition, architecture, engineering, social and earth sciences. The Institute also sponsors the quarterly international journal *Disasters*. Whereas most American hazard geographers are primarily social scientists, apart from those involved in the Disaster Research Unit, most British hazard geographers are physical scientists trained in the analysis of geomorphic, hydrologic or meterological processes.[28]

Geographers who participated in the work of the International Geographical Union Commission on Man and Environment played a formative role in Australian hazard research. They supplied a majority of the papers delivered at a nationwide Hazard Research Symposium in 1976.[29] This brought together, for the first time,

---

[27]. "Seventy-Five Years of American Geography," special issue of *Annals of the Association of American Geographers* 69 (March 1979): 185 pp.

[28]. Clark, loc. cit.; Gregory and Williams, loc. cit.; Smith and Tobin, op. cit.; Perry, op. cit.; Parking and Harding, loc. cit.; Dennis J. Parker and Donald M. Harding, "Planning for Urban Floods," *Disasters* 2 (1978): 47-57; Dennis J. Parker and E.C. Penning-Rowsell, "Specialist Hazard Mapping: The Water Authorities' Land Drainage Surveys," *Area* 13 (2) (1981): 97-108; A.M. Duncan, D.K. Chester, and J.E. Guest, "Mount Etna Volcano: Environmental Impact and Problems of Volcano Prediction," *Geographical Journal* 147 (July 1981): 164-178.

representatives of all hazard interest groups in Australia and it paved the way for a more intensive research program focusing on natural hazards in the northern half of the country.

As hazard geographers broadened the scope of their work during the 1970s, the range of contacts with researchers from other fields also expanded. Existing ties to economists and psychologists were consolidated and new cooperative hazard research links were established with sociologists, anthropologists, archaeologists and political scientists. As a result there was growing recognition of the common interests that united Hazard Researchers with specialists in overlapping fields like emergency planning; stress management; disaster research; risk management; decision theory; and human ecology. Some researchers began to call for the synthesis or merging of related fields.[30] Exposure to the work of such groups strengthened Hazard Research by subjecting its assumptions and findings to critical examination and by providing it with new theoretical and methodological insights. However, as a byproduct, boundary disputes and other conflicts also occurred between hazard researchers and scholars from competing research traditions. Not the least among these were geographers who espoused alternative paradigms of geographical inquiry.

Issues and Disputes in Contemporary Hazard Geography

In recent years three main theoretical issues have been raised by critics of Hazard Research. These concern: (1) the role of choice; (2) the salience of hazard; and (3) the character of collective behavior. All three issues have clear implications for Hazard Perception Studies.

Many Hazard geographers advocate the desirability of expanding the range of choice available to populations at risk. This is generally achieved by identifying and evaluating known adjustments to specific hazards, and by suggesting ways to improve prospects for their adoption. Some humanistic geographers have suggested an alternative approach which emphasizes " . . . enlarging [the] horizons of consciousness [of affected groups] to the point where both the articulation of alternatives and the choice of direction could be theirs."[31] Although there is reason to believe that many hazards geographers are sympathetic to Buttimer's goal--and hence

---

29. Heathcote and Thom, op. cit.

30. John R. Gold, "The Environment as Hazard," in ed. I. Burton, R.W. Kates and G.F. White *Disasters* 3 (1979): 328-329.

31. Sister Annette Buttimer, *Values in Geography,* Resource Paper no. 24, (Washington, D.C.: Association of American Geographers, 1974), p. 29.

that her criticism is misplaced--this presumed divergence of orientation suggests a need for a more critical look at conceptions of choice which underpin hazard research.

The purposes of offering people enhanced opportunities to choose have not received significant attention from Hazard Researchers. Nor have relationships between choice and accountability or legal and moral responsibilities. Although decision scientists are beginning to uncover some of the limitations on our capacity for choice, much more remains to be learned.[32] Thus, Hazard Perception research has yet to grapple with important aspects of the rules of action that govern hazard choices. Furthermore, perception researchers face serious difficulties in answering such questions. Willingness to choose, autonomy of choice, capacity for choice, and mechanisms for mediating expediencies and moral imperatives, possess ethical dimensions that lie beyond behavioral science but cannot be ignored by hazard scholars.

Rossi and his co-workers at the University of Massachusetts have questioned whether natural hazards are salient human problems.[33] They judge that much natural hazard research may deal with inconsequential questions and that associated policy recommendations probably seek unrealistic and unobtainable goals that are more appropriate for the resolution of high salience issues. This conclusion is based on an analysis of population, housing and economic data from 1960 to 1970 census tracts in communities affected by natural disasters. Rossi's methods and findings have generated heated controversy and strong rebuttals,[34] but the issue of salience is of continuing significance. Apart from one early study, the literature of Hazard Perception has not explored this topic.[35]

Most critics of Hazard Research do not contest the relevance or salience of the field. They focus instead on contingent theoretical or pardigmatic issues. The chief complaint is that

---

32. Slovic, Fischhoff, and Lichtenstein, loc. cit.; P. Slovic, B. Fischhoff, and Sarah Lichtenstein, in *Societal Risk Assessment*: *How Safe is Safe Enough?* ed. R. Schwing and W.A. Albers, Jr. (New York: Plenum, 1980), pp. 181-214; P. Slovic, B. Fischhoff, and Sarah Lichtenstein, "Perceived Risk: Psychological Factors and Social Implications," *Proceedings of the Royal Society of London* A376 (1981): 17-34.

33. James D. Wright, Peter H. Rossi, Sonia R. Wright and Eleanor Weber-Burdin, *After the Clean-Up*: *Long Range Effects of Natural Disasters*, Contemporary Evaluation Research no. 2 (Beverly Hills: Sage Publications, 1979); James D. Wright, and Peter H. Rossi, eds., *Social Science and Natural Hazards* (Cambridge: Abt Books, 1981).

34. Wright and Rossi, op. cit., pp. 171-177.

35. Stephen Golant and Ian Burton, *The Meaning of a Hazard--Application of the Semantic Differential*, Natural Hazard Research Working Paper no. 7 (Toronto: University of Toronto, Department of Geography, 1969).

Hazard Research possesses a meagre theoretical foundation, especially in reference to the interpretation of collective behavior.[36] Existing theory emphasizes individual perception decisions about hazard mitigation in the context of a man environment systems model with a high level of generality. Despite occasional attempts to investigate differences between individual and collective coping there has been little development of collective decision theory. Critics imply that present individual perception oriented approaches are therefore misleading because collective decision processes dominate the selection of adjustments to hazard.

At least four or five alternative viewpoints have been suggested. (A) Sociologists identified with the Disaster Research Center at Ohio State University have argued that groups are the principal unit of description and analysis in studies of how humans cope with extreme events.[37] They base their judgment on Fritz's observation that " . . . at least at the practical and operational level, the effectiveness and efficiency of disaster response is dependent more on the vitality of the emergency organizations involved in the crisis than it is on the psychological state or readiness of individual victims."[38] As a consequence, disaster researchers have accumulated a large body of material about organizational behavior and emergent groups.[39] However, most of this work focuses on the immediate post-disaster period and it has limited value for explaining the broad span of hazards appraisals and responses employed by populations at risk before a disaster occurs or in the period of long-term rehabilitation.

(B) A second viewpoint on collective behavior has been advanced by cultural and human ecologists.[40] By applying natural

---

36. R.C. Golledge, "Misconceptions, Misinterpretations, and Misperceptions of Behavioral Approach in Human Geography," *Environment and Planning* Series A, 13 no. 11, (1982), pp. 1325-1344; Dennis S. Mileti, Thomas E. Drabek, and J. Eugene Haas, *Human Systems in Extreme Environments,* p. 99; E.L. Quarantelli and Russell R. Dynes, "Response to Social Crises and Disaster," *Annual Review of Sociology* 3 (1977): 23-49; Perry, op. cit.; Julian Wolpert, "The Dignity of Risk," pp. 391-401; William I. Torrey, "Hazards, Hazes and Holes: A Critique of *The Environment as Hazard* and General Reflection on Disaster Research," *Canadian Geographer* 24 (Autumn 1981): 286-289; William I. Torry, "Anthropological Studies In Hazardous Environments: Past Trends and New Horizons," *Current Anthropology* 20 (September 1979): 43-52.

37. E.L. Quarantelli and Russell R. Dynes, "Response to Social Crisis and Disaster," p. 30.

38. Russell R. Dynes, "The Comparative Study of Disaster: A Social Organizational Approach," *Mass Emergencies* 1 (1975): 22-31.

39. E.L. Quarantelli, ed., *Disasters: Theory and Research* (Beverly Hills: Sage Publications, 1978); Dynes, loc. cit., pp. 21-31.

40. Eric Waddell, "The Hazards of Scientism: A Review Article," *Human Ecology* 5 (1977): 69-76; Gilbert F. White, "Natural Hazards and the Third World--A Reply," *Human Ecology* 6 (June 1978): 229-231; Kenneth Hewitt, *The Environment as Hazard* by I. Burton, R.W. Kates and G.F. White in *Annals of the Association of*

ecological systems theory to human populations, it is reasoned that human groups respond to stress in a two stage manner.[41] Initially, they seek to establish homeostasis or equilibrium by directing resources to cope with threatening hazards. If this fails, they turn to maintaining resilience by safeguarding vital system operations that are needed for survival. Drawing on these ideas, Morren has outlined a hierarchy of response processes which serve one or both functions.[42] Other anthropologists have offered related models of group behavior that highlight environmental, economic or social constraints on freedom of action, and the primacy of traditional, taken-for-granted coping systems.[43]

(C) Scholars associated with Bradford University's Disaster Research Unit champion a third interpretation of collective behavior in the face of hazard.[44] This stresses the existence of constraints on individual freedom of choice to adopt hazardous adjustments. Particularly in poor nations, traditional hazard coping strategies that once incorporated the bulk of populations at risk are seen as being undermined and replaced by bureaucratized systems which are only available, and useful, to elite groups. Thus, Wisner's study of six hundred farm families in Eastern Kenya revealed that selection of adjustments to drought is related less to perceptions of hazard than to economic and environmental constraints which severely restrict ability to choose among alternative strategies.[45] While a relatively few well-off farmers benefit from increased

---

American Geographers 70 (June 1980): 306-311.

41. A.P. Vayda and B.J. MacKay, "New Directions in Ecology and Ecological Anthropology," *Annual Review of Anthropology* 4 (1975): 293-306.

42. George Morren, "The Rural Ecology of the British Drought of 1975-1976," *Human Ecology* 8 (1980): 33-63.

43. Francesco Battisti, "Thresholds of Security in Different Societies," *Disasters* 4 (October 1980): 101-105; John Brine, "After Cyclone Ted on Mornington Island: The Accumulation of Physical and Social Impacts on a Remote Australian Aboriginal Community," *Disasters* 4 (October 1980): 3-10; Louis Evan Griveth, "Geographical Location, Climate and Weather, and Magic: Aspects of Agricultural Success in the Eastern Kalahari, Botswana," *Social Social Science Information* 20 (1981): 509-536; David Guillet, "Surplus Extraction, Risk Extraction and Economic Change Among Peruvian Peasants," *Journal of Development Studies* 18 (October 1981): 3-24; Richard Hogg, "Pastoralism and Impoverishment: The Case of the Isiolo Boran of Northern Kenya," *Disasters* 4 (1980): 299-310; Stephen Nugent, "Amazonia: Ecosystem and Social System," *Man* 16 (March 1981): 62-74; Jeremy Swift, "Sahelian Pastoralists: Under-Development, Desertification and Famine," *Annual Review of Anthropology* 6 (1977): 457-478; Hector Luis Morales Zavala, "Rural Development, Science and Political Decision-Making: Diverging or Comerging Tendencies?," *Impact of Science on Society* 30 (July/September 1980): 167-178.

44. P. O'Keefe and K. Westgate, "Preventive Planning for Disasters," *Long Range Planning* 10 (June 1977): 25-29; K. Westgate and P. O'Keefe, *Some Definitions of Disaster*, Disaster Research Unit Occasional Paper no. 4 (Bradford: University of Bradford, 1976); B. Wisner, P. O'Keefe, and K. Westgate, "Global Systems and Local Disasters: The Untapped Power of Peoples' Science," *Disasters* 1 (1977): 47-57.

45. Benjamin Wisner, *An Overview of Drought in Kenya*, Natural Hazard Research Working Paper no. 30 (Boulder: University of Boulder, Institute of Behavioral Science, 1976).

choices made available by governmental programs and modern technology, this is at the expense of the majority which suffers further restricted opportunities. This variant of development and dependency theory regards increasing "marginalization" of the poor and politically weak in hazardous locations as a logical outcome of systematic economic development processes that have their origin in competition within and between capitalism, socialism and other global economic systems. Other researchers have come to similar, if less deterministic, conclusions.46 For example, Sonnenfeld argues that the process of modernization encourages reductions in risk-taking behavior among Alaskan Eskimos and weakens social relationships that provided means for overcoming the deficiencies of a subsistence lifestyle.47

(D) Political scientists have begun to focus attention on the type of hazard adjustment decision at issue rather than the process of human response. They suggest that such collective (public) decisions may best be thought of as alternative outputs from political allocation systems.48 Drawing on work by Lowi, they argue that different forms of "distributive politics" predictably lead to different degrees of public acceptance or rejection of adjustments.

(E) Finally, historians and archaeologists have stressed both the evolving and interacting permutations of circumstances which give rise to unique patterns of hazard and to the persistence of long-term conditions that structure or limit societal responses to major dislocations.49 The archaeological record illustrates the contrasting consequences of similar sized volcanic eruptions.50 Likewise, Bosworth explains the slow recovery of Sicily from the Messina earthquake as an outgrowth of great themes of Italian

---

46. K. Hewitt, "Earthquake Hazards in the Mountains," *Natural History* (May 1976): 30-37; Susan E. Jeffery, "Our Usual Landslide," *Ubiquitous Hazard and Socioeconomic Causes of Natural Disaster in Indonesia,* Natuaral Hazard Research Working Paper no. 40 (Boulder: University of Colorado, Institute of Behavioral Science, 1981).

47. J. Sonnenfeld, "Social Stress in Extreme and Remote Environments," unpublished paper presented at the Annual Meeting of the Association of American Geographers, Milwaukee 1975.

48. Richard Stuart Olson and Douglas C. Nelson, "Public Policy Analysis and Hazards Research: Natural Complements," *The Social Science Journal* (January 1982): 89-103.

49. Karl W. Butzer, "Cultural Adaptation: Exploration of an Idea," unpublished paper delivered at the Annual Meeting of the Association of American Geographers, Louisville, 1980; Karl W. Butzer, "Civilizations: Organisms or Systems?," *American Scientist* 68 (September/October 1980): 517-523; Eric L. Jones, "Disaster Management and Resource Saving in Europe, 1400-1800," in *Natural Resources in European History,* ed. by Antoni Maczak and William N. Parker, RFF Research Paper no. R-13 (Washington, D.C.: Resources for the Future Inc., 1978): 114-138.

50. Payson D. Sheets, *Archaeological Studies of Disaster: Their Range and Value,* Natural Hazard Research Working Paper no. 38 (Boulder: University of Colorado, Institute of Behavioral Science, 1980); P.D. Sheets and D.K. Grayson, eds., *Volcanic Activity and Human Ecology* (New York: Academic Press, 1979).

history which expressed themselves as an "... intricate fabric of despair and alienation, of that hopelessness which is the real basis of the Southern Question."[51]

To the extent that: (1) human responses to hazard are severely constrained by lack of alternative choices; (2) hazards are low salience issues; and (3) collective behavior is a potent influence on the selection of adjustments, the importance of individual variations in the perception of risks and adjustments is likely to be reduced. Inasmuch as these were three popular research topics in the 1970s, it is not surprising that the relative pace of Hazard Perception Studies somewhat slackened during the decade. The published research literature in Hazard Research has grown prodigiously in the past decade but Hazard Perception Studies have not kept pace with this general expansion. It is true that many Hazard Research publications mention the concept of perception but studies devoted primarily to that topic represent a much smaller portion of the field. Indeed, some major publications on social science aspects of hazards barely touch on the subject.[52] Aside from the voluminous work on risk assessment by Slovic, Fischoff and Lichtenstein and their associates, and a growing number of studies dealing with definitional problems that bear on hazard management (e.g., "desertification," "water conservation")[53] only thirty to forty significant articles on hazard perception have appeared in English language journals since 1972. This represents no more than 15% of the total output of articles on Hazard Research during the same period.

Hazard Perception Studies

Mileti has summarized the findings of Hazard Perception Studies in the mid-1970s.[54] Following White, Slovic and others, he suggested that Risk (Hazard) Adjustment is a four step process involving: (1) assessment of the probability of a natural extreme; (2) review of alternative adjustments available; (3) evaluation of the impacts of alternative adjustments; and (4) choice of adjustment. Together with (a) characteristics of the hazard manager

---

51. R.J.B. Bosworth, "The Messina Earthquake of 28 December, 1908," *European Studies Review* 11 (1981): 189-206.

52. Harold D. Foster, *Disaster Planning*; Gilbert F. White and J. Eugene Haas, *Assessment of Research on Natural Hazards*.

53. D.A. Carder, "Desertification in Australia--A Muddled Concept," *Search* 12 (July 1980): 217-221; D.D. Baumann, J.J. Boland and J.H. Sims, "Water Conservation: The Struggle over Definition," unpublished paper (Carbondale: Department of Geography, Southern Illinois University, 1981).

54. Dennis S. Mileti, "Human Adjustment to the Risk of Environmental Extremes," pp. 327-347.

(social unit), and (b) the influence of other groups and institutions (inter-system incentives), (c) risk perception is considered to be a key variable in the adjustment process. (Fig. 3.2) Risk perception is in turn viewed as a function of seven other variables. (Fig. 3.3) In Mileti's words:

> Despite faults in human cognition of risk, the probability of risk-mitigating adjustment increases as a positive function of risk perception through the mediating effect that perceived risk has on the variables of image of damage and perceived benefits of such adjustment . . . Image of damage is what social units think will happen to themselves, possessions and community were an environmental extreme to occur; it has a positive effect on both perceived benefits of risk-mitigation policy and on risk-mitigation adjustment. The more potential damage imputed on the basis of risk, the more likely a social unit will adjust to that risk. Perceived benefits, positively affect the probability of risk-mitigating adjustment to the extent that anticipated benefits are worth the costs of policy implementation.55

Mileti's model is largely based on work carried out in the 1950s, 1960s and early 1970s. How well do the findings of more recent Hazard Perception Studies fit his formulation? In fact, most of the perception components of his model are essentially untested. Recent work has highlighted only two: (1) links between experience of hazard, awareness and risk perception (i.e., hazard awareness studies); and (2) propensities to deny or distort risk assessments (i.e., perceptual bias studies).

Hazard Awareness Studies

These are generally broad-based data gathering investigations undertaken for the purposes of improving public policy responses to specific hazards. They usually define the information base from which judgements of hazard are made and estimate the impact of inaccurate perceptions on adjustment decisions. Useful information about images of hazards and adjustments is thereby gained. These data are interpreted within, and help to test, the general systems learning model of hazard adjustment developed by Kates and others.56 However, little new conceptual or theoretical ground pertaining to perception processes, is broken. Several examples illustrate these points.

In a study to determine the longevity of previous hazard experiences, Hanson et al. concluded that most residents of Flint, Michigan clearly remembered the devastating 1953 tornado.57 Those who lived near the tornado's path possessed the most heightened

---

55. Mileti, loc. cit., p. 337.

56. Robert W. Kates, *Natural Hazard in Human Ecological Perspective*.

57. Perry O. Hanson, John D. Vitek, and Susan Hanson, "Awareness of Tornadoes: the Importance of an Historic Event;" Susan Hanson, John Vitek, and Perry Hanson, "Natural Disaster: Long Range Impact on Human Response to Future Disaster Threats," *Environment and Behavior* 11 (June 1979): 278-283.

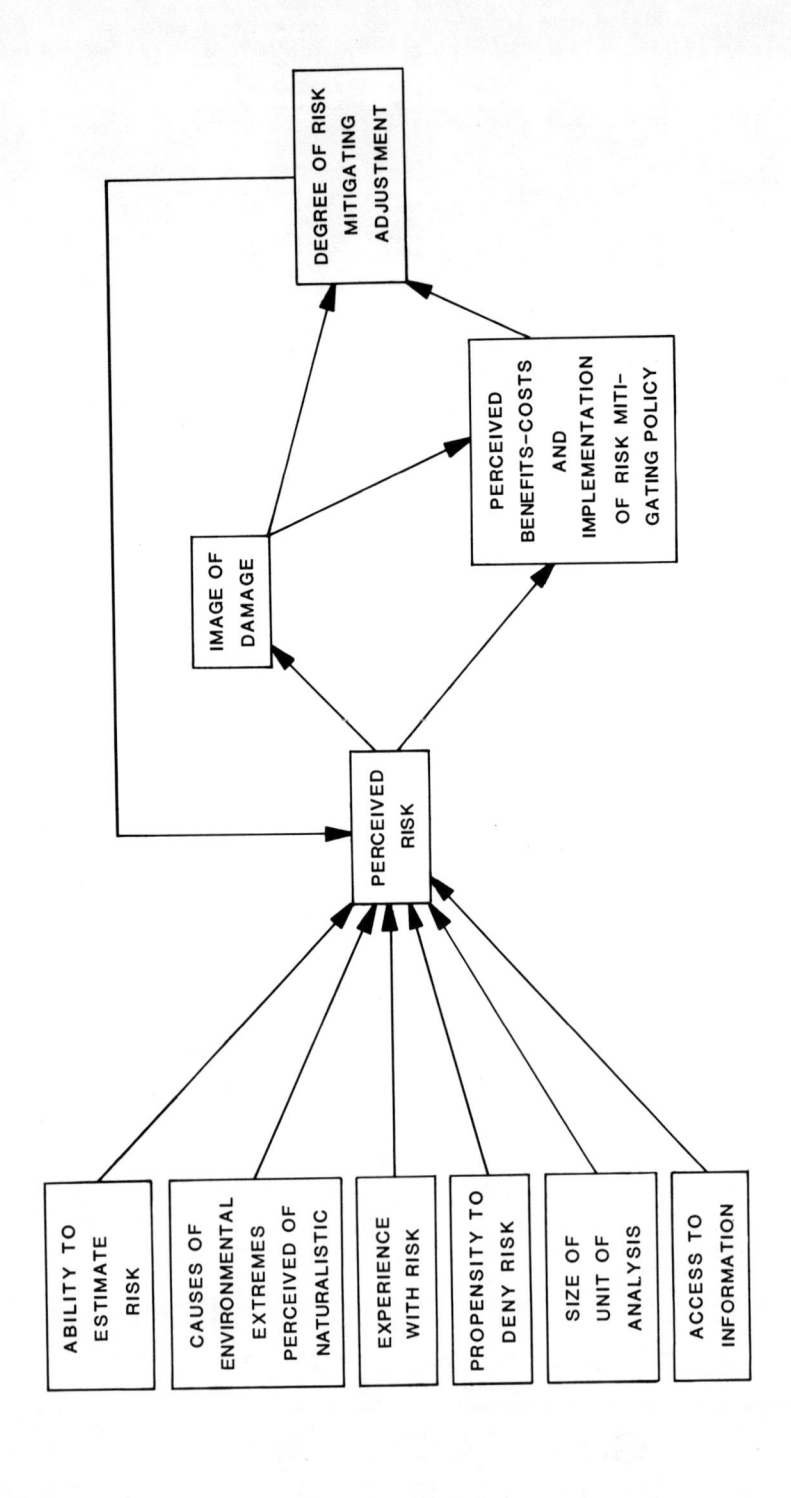

Figure 3.3 Causes and Consequences of Risk Perception (After Mileti, 1980)

awareness and were most likely to take the most effective protective actions in the event of another tornado.

Smith and Tobin employed a questionnaire to gather information about public awareness of floods and adjustments in Northwest England.[58] This revealed that few people were aware of existing warning schemes. It also demonstrated that personal experience is the principal stimulus to recognition of flood hazard and that younger residents place more faith in public flood control schemes while seriously overestimating the efficacy of private remedial actions.

Jackson followed up an inconclusive study of earthquake perception in San Francisco by modifying a questionnaire originally used in the 1971 International Geographical Union Commission on Man and Environment studies,[59] and by expanding his sample of respondents to include people living in Anchorage, Vancouver, Victoria and Los Angeles.[60] His results indicate that earthquake perceptions and responses are strongly influenced by previous experience of the hazard; its salience, respondents' emotional states; and individual beliefs about personal responsibility for safety.

Cross determined that hurricane awareness is high in the Florida Keys but that most residents regard hurricanes as low salience problems.[61] Although residents who arrived during the stormy 1960s paid more attention to hurricane safety when selecting homes, few people chose elevated houses primarily for their storm safety features. Liverman judged that there was little consistency among perceptions of the causes, nature and responses to drought in Western Canada.[62]

Following Saarinen and others, Eastwood and Carter used Semantic Differential scales and Thematic Apperception tests to explore psychological attributes of recreationalists in a fragile dune complex.[63] They found that beach users and dune users recorded

---

58. K. Smith and Graham Tobin, *Human Adjustment to the Flood Hazard*.

59. Gilbert F. White, ed., *Natural Hazards*, pp. 6-9.

60. E.L. Jackson and T. Mukerjee, "Human Adjustment to the Earthquake Hazard of San Francisco, California," in *Natural Hazards: Local, National, Global*, ed. by G.F. White (New York: Oxford University Press, 1976), pp. 160-166; Edgar L. Jackson, "Response to Earthquake Hazard: The West Coast of North America," *Environment and Behavior* 13 (July 1981): 387-416.

61. John A. Cross, "Residents' Concerns about the Hurricane Hazard within the Lower Florida Keys," in *Hurricanes and Coastal Storms*, ed. by Earl J. Baker, Report no. 33 (Talahassee: Florida Sea Grant College, 1980), pp. 156-161.

62. Diana M. Liverman, *Perception and Communication in Government Response to the 1977 Drought in Western Canada*, Emergency and Risk Research Working Paper no. 1 (Toronto: University of Toronto, Institute of Environmental Studies, 1978).

63. D.A. Eastwood and R.W.G. Carter, "The Irish Dune Consumer," *Journal of*

contrasting scores on Semantic scales. However, very few respondents recognized links between heavy use of dunes and severe environmental degradation. Hence, they concluded that--unlike more dramatic natural hazards--dune degradation is not readily perceived by dune users because " . . . the rate of geomorphological and ecological change is both subtle and largely out-of-phase with maximum visitor pressure."[64]

These and similar studies,[65] clearly substantiate a basic Natural Hazard Research thesis; namely, that populations at risk and public hazard managers often perceive hazards in ways that are in variance with actual characteristics and thus may be led to make inappropriate adjustment decisions. These studies also tend to support the assertion that experience of hazard heightens risk perception. However, the evidence is not unanimous. Drawing on studies in a wide range of fields, Baumann and Sims cite contrary examples.[66] In this way they are supported by James and Wenger's surveys of beliefs about post-disaster behavior. Analysis of disaster-prone and disaster-free communities reveals that erroneous beliefs dominate everywhere.[67] Most people hold stereotypes of disaster behavior such as the existence of widespread panic, looting and helpless victims. Perhaps the apparent contradictions are easily explained? Images of the consequences of past disasters may stimulate heightened awareness whereas details of those ongoing events may be less clearly remembered or less willingly retained.

In some instances, experience also seems to have a direct bearing on at least some forms of adjustment behavior. People who have experienced hazard are more likely to adopt engineering structures, emergency production measures and land use controls.[68]

Leisure Research 13 (1981): 273-381.

64. Eastwood and Carter, loc. cit., p. 281.

65. Sally Davenport, *Human Responses to Hurricanes in Texas--Two Studies*, Natural Hazard Research Working Paper no. 34 (Boulder: University of Colorado, Institute of Behavioral Science, 1978); David E. Kromm, "Response to Air Pollution in Ljubljana, Yugoslavia," *Annals of the Association of American Geographers* 63 (June 1973): 208-217; David E. Kromm, F. Probald, and G. Wall, "An International Comparison of Response to Air Pollution," *Journal of Environmental Management* 1 (October 1973): 363-376; H.F. McPherson and T.F. Saarinen, "Flood Plain Dwellers Perception of the Flood Hazard in Tucson, Arizona," *Annals of Regional Science* 11 (1977): 25-40; John Oliver, *Natural Hazards: Response and Planning in Tropical Queensland*, Natural Hazards Research Working Paper no. 33 (Boulder: University of Colorado, Institute of Behavioral Science, 1978); White, op. cit.

66. Duane D. Baumann and John H. Sims, "Educational Programs and Human Response to Natural Hazards," unpublished paper (Carbondale: Department of Geography, Southern Illinois University, 1981).

67. Thomas F. James and Dennis E. Wenger, "Public Perceptions of Disasters--Related Behavior," in *Hurricanes and Coastal Storms,* ed. by Earl J. Baker, Report no. 33 (Talahassee: Florida Sea Grant College, 1980), pp. 162-168; Dennis E. Wenger, James D. Dykes, Thomas D. Sebok, and John L. Neff, "It's a Matter of Myths: An Empirical Examination of Individual Insights into Disaster Response," *Mass Emergencies* 1 (1975): 33-46.

However, the burden of opposing evidence is also heavy. As Baumann and Sims assert:

> . . . the link between experience and behavior leaves much to be desired. It is not a clean and clear one. Indeed, sometimes it does not exist, and in other instances, and perhaps even more disturbing, the relationship though there, results not in protection but in harm.[69]

This is particularly clear in studies of public information programs and warning responses. Despite numerous findings that lay citizens and public officials desire improved information about threatening hazards[70] " . . . neither increased awareness nor behavior change can be assumed just because information is made available to people."[71]

The optimum conditions for responding to hazard warnings are well known.[72] However, the number and complexity of necessary actions, and the range of actors, are such that it is extremely difficult to predict actual responses to a given hazard warning. For example, it is generally agreed that people respond favorably to warnings when they: (1) define the threat as real; (2) perceive significant personal risks; and (3) possess a personal response strategy. Yet, the existence of these conditions depends on other factors such as: the type, content, timing and source of warning messages; previous hazard experiences; levels of social reinforcement; knowledge of preparedness plans; and personality

---

68. Earl J. Baker, "Public Attitudes Toward Hazard Zone Controls," *Journal of the American Institute of Planners* 43 (October 1977): 401-408; Earl J. Baker and D.J. Patton, "Attitudes toward Hurricane Hazards on the Gulf Coast," in *Natural Hazards: Local, National, Global*, ed. by Gilbert F. White (New York: Oxford University Press, 1974): 30-36; Martyn J. Bowden et al., *Hurricanes in Paradise: Perception and Reality of Hurricane Hazard in the Virgin Islands* (St. Thomas: Island Resources Foundation, 1974); Charlene C. Levy, "Apathy and Inexperience: A Study of Hurricane Hazard Perception," PPA-28, paper delivered at the Natural Hazards Research Applications Workshop, Boulder, Colorado, 1979; Marvin Waterstone, *Hazard Mitigation Behavior of Urban Flood Plain Residents*, Natural Hazards Research Working Paper no. 35 (Boulder: University of Colorado, Institute of Behavioral Science, 1979).

69. Baumann and Sims, loc. cit., p. 16.

70. Kent Barnes, James Brosius, Susan C. Cutter, and James K. Mitchell, *Responses of Impacted Populations to the Three Mile Island Nuclear Accident*; Davenport, op. cit.; Denise Haller Paz, Ralph H. Turner and Joanne M. Nigg, "Social Response to Earthquake Prediction Information: Attitudes toward Prediction Capability and Government Action" CRP-7, presentation at the Natural Hazards Workshop, Boulder, 1980; Thomas F. Saarinen and Harold J. McPherson, *Notices, Watches and Warnings: An Appraisal of the U.S.G.S.'s Warning Systems with a Case Study from Kodiak, Alaska*, Natural Hazards Research Working Paper no. 42 (Boulder: University of Colorado, Institute of Behavioral Science, 1981); John H. Sorensen, *Emergency Response to Mount St. Helens' Eruption: March 20 to April 10, 1980*, Natural Hazards Research Working Paper no. 43 (Boulder: University of Colorado, Institute of Behavioral Science, 1981); Donald Zeigler, Stanley D. Bruun and James H. Johnson, Jr., "Evacuation from a Nuclear Technological Disaster," *Geographical Review* 71, no. 1 (January 1981), pp. 1-16.

71. Thomas F. Saarinen et al., "How can we Achieve Wider and More Effective Dissemination of Warnings?," SS-1, record of Special Session, Natural Hazards Workshop, Boulder, 1981. See also: Earl J. Baker, ed., *Hurricane and Coastal Storms*, Report no. 33 (Talahassee: Florida Sea Grant College, 1980); Baumann and Sims, loc. cit.; John P. Clark and T. Michael Carter, "Response to Hurricane Warnings as a Process: Determinants of Household Behavior," in Baker, op. cit., pp. 19-24; Gerold O. Windham, et al., "Reactions to Storm Threat During Hurricane Eloise," Report no. 51 (Jackson: Mississippi State University, Social Science Research Center, 1977); Risa Palm, "Public Response to Earthquake Hazard

variables. Moreover, these factors can be disaggregated still further. Thus, Ruch has examined the effects of different types of hazard awareness messages on perception of hurricane dangers in Galveston, Texas.73 He concluded that printed brochures increased knowledge of hurricanes but reduced cautionary attitudes. Early messages in a series of television bulletins conferred no appreciably heightened awareness but later messages did. Early radio bulletins depressed awareness of hazard whereas later messages heightened awareness. Clearly, in the case of responses to hazard warning systems, perception is but one element in a complex decision system. Perhaps hazard information and warning systems have least influence on human behavior when individuals are contemplating large economic and psychic investments (e.g., house purchases; evacuation).74 Conversely, they may be most effective in bringing about behavior changes when proposed adjustments require: (a) relatively modest outlays of time, effort and money (e.g. emergency protection); (b) minimal personal costs (e.g. insurance); or (c) slow to mature, collective actions (e.g. major structural protection projects, land use controls).

Further light on links between experience, awareness and adjustment is provided by studies of newly emerging threats that were previously unknown to populations at risk. For example, researchers at the Batelle Human Affairs Center used a brief telephone questionnaire to assess hazard perceptions of people living near Mt. St. Helens before its catastrophic May 1980 eruption.75 Information was solicited about respondents' risk estimates and beliefs concerning volcano hazards, their information sources, and their proposed responses. Though frequently informed by the mass media about the existence of a threatening volcano, most residents viewed an eruption as a low probability event not requiring any serious preparedness planning. In a related study,

---

Information," *Annals of the Association of American Geographers* 71 (September 1981): 389-399.

72. Dennis S. Mileti, Thomas E. Drabek, and J. Eugene Haas, *Human Systems in Extreme Environments*; Ronald William Perry, Michael K. Lindell, and Marjorie R. Greene, *The Implications of Natural Hazard Evaluation Warning Studies for Crisis Relocation Planning*, B-HARC-411-035 (Seattle: Battelle Human Affairs Research Center, 1980); E.L. Quarantelli, *Evacuation Behavior and Problems: Findings and Implications from the Research Literature,* Report 7-80 (Columbus: Ohio State University, Disaster Research Center, 1980).

73. Larry Christensen and Carlton E. Ruch, "The Effect of Social Influence on Response to Hurricane Warnings," *Disasters* 4 (1980): 205-210; Carlton Ruch, "Awareness Program Component Assessment," in Baker, op. cit., pp. 143-149.

74. Palm, loc. cit., 389-399; John A. Cross, "Residents' concerns about the Hurricane Hazard within the Lower Florida Keys," in Baker, op. cit., pp. 156-161.

75. M. Greene, R. Perry and M. Lindell, "The March 1980 Eruptions of Mt. St. Helens: Citizen Perceptions of Volcano Threat," *Disasters* 5 (1981): 49-66.

Sorensen largely confirmed these findings.[76] He also noted that local public officials, responsible for emergency management, possessed only vague ideas about the nature and consequences of an eruption. Visitors to the area looked forward to a major dramatic eruption whereas locals hoped for a series of unspectacular small events.

A study of responses to a nuclear accident revealed different findings. One third of those living near Three Mile Island quickly left the area; one third waited for further evacuation clues; and one third reported no intention of leaving.[77] Likewise, 70% of persons interviewed in Elizabeth, New Jersey reported that they felt in danger from a major toxic chemical explosion and fire. Fifty percent took immediate precautions to protect their lives. However, no one left the area and only 7% made plans to evacuate.[78] None of these populations had ever experienced similar types of problems. These findings suggest that lack of experience may limit, but does not necessarily prevent, the adoption of protective adjustments. Unknown or dreaded hazards may call forth active responses on a large scale without the intervention of educational campaigns and specific hazard warning programs.[79] Of course, such campaigns and programs may also increase the level of responsiveness.

Perceptual Process Studies: A Note on Techniques

By 1972 hazard researchers were already experimenting with the application of several psychological testing techniques and procedures. These included Thematic Apperception Tests; Achievement Scores; Semantic Differential scales; Sentence Completion tests; Rosenzweig Picture Frustration tests; and various measures of stimulus avoidance and response. Several of these procedures evidently elicited little enthusiasm and soon fell into disuse. In contrast, during the 1970s, TATs and Sentence Completion tests continued as widely used research tools.[80] New techniques like

---

76. Sorensen, op. cit.

77. Barnes, Brosius, Cutter, and Mitchell, op. cit.

78. Susan Cutter, Stephen Decter, James Brosius, and Charles Kelly, *Institutional and Individual Responses to Toxic Chemical Fires: The Chemical Control Corporation Fire, April 21, 1980, Elizabeth, New Jersey*, Rutgers University Department of Geography Discussion Papers, no. 17 (New Brunswick: Rutgers University Department of Geography, 1980).

79. P. Slovic, B. Fischhoff, and Sarah Leichtenstein, "Perceived Risk: Psychological Factors and Social Implications," pp. 17-34.

80. Duane D. Baumann and John H. Sims, "Flood Insurance: Some Determinants of Adoption," *Economic Geography* 54 (July 1978): 189-196; D.A. Eastwood and R.W.G. Carter, "The Irish Dune Consumer," pp. 273-281; James K. Mitchell, *Community Response to Coastal Erosion: Individual and Collective Adjustments to Hazard on the Atlantic Shore*, University of Chicago, Department of Geography Research Papers, no. 156 (Chicago: University of Chicago, Department of Geography, 1974), pp. 81-88; John Sims and Duane D. Baumann, "The Tornado Threat: Coping Styles

cognitive mapping;[81] personality tests;[82] and measures of sensation seeking[83] and repression sensitization[84] were added. These devices were generally used in the context of studies which explored either locus of control or cognitive dissonance interpretations of hazard perception.

Tests of Perception Theory

Early hazard perception studies frequently interpreted their finds in terms of cognitive dissonance theory.[85] They explained hazard zone residents' or users' denials of risk as logical outcomes of the need to continue to occupy and use these areas. Shippee and his colleagues tested an extension of this proposition which included distance from a hazard source as an independent variable.[86] They found that the perception of hazard was enhanced among residents who considered themselves to live on and beyond the outer fringes of the risk zone and who lacked visual reinforcement clues about the hazard. However, no evidence was found to support a second hypothesis that people living near a hazard zone tend to deny the threat. Obviously, distance from hazard is not a particularly useful measure for many hazard studies (e.g. drought) and one test is not sufficient to discredit the early findings. Nonetheless, scattered findings from recent unrelated work tend to add weight to Shippee's conclusions. For example, evacuation journeys of residents around Three Mile Island were proportional in length to the distances of their homes from the reactor.[87] A lower percentage of occupants at distant sites left the area but they went much

---

North and South," *Science* 176 (1972): 1386-1392.

81. Levy, loc. cit.

82. Eastwood and Carter, loc. cit.

83. M.R. Schiff, "Hazard Adjustment, Locus of Control, and Sensation Seeking: Some Null Findings," *Environment and Behavior* 9 (1977): 233-254.

84. Paul Simpson-Housley, *Locus of Control, Repression-Sensitization and Perception of Earthquake Hazard,* Natural Hazard Working Paper no. 36 (Boulder: University of Colorado, Institute of Behavioral Science, 1978).

85. Robert L.A. Adams, "Uncertainty in Nature, Cognitive Dissonance and the Perceptual Distortion of Environmental Information: Weather Forecasts and New England Beach Trip Decisions," *Economic Geography* 49 (October 1973): 287-297; Ian Burton, "Cultural and Personality Variables in the Perception of Natural Hazards," in *Environment and the Social Sciences: Perspectives and Applications,* ed. by J.F. Wohlwill and H. Carson (Washington, D.C.: American Psychological Association, 1972); Stephen Golant and Ian Burton, *The Meaning of Hazard* E.L. Jackson and T. Mukerjee, "Human Adjustment to the Earthquake Hazard of San Francisco, California;" J.M. Hewings, *Water Quality and the Hazard to Health: Placarding Public Beaches,* National Hazard Research Working Paper no. 3 (Toronto: University of Toronto, Department of Geography, 1968).

86. Glenn Shippee, Jeffrey Burroughs, and Stuart Wakefield, "Dissonance Theory Revisited: Perception of Environmental Hazards in Residential Areas," *Environment and Behavior* 12 (March 1980): 33-51.

87. Barnes, Brosius, Cutter, and Mitchell, op. cit.; Zeigler, Brunn and Johnson, loc. cit.

further to seek safety. This suggests though threat enhancement may occur at sites outside an immediate hazard zone, affected individuals face increasingly strong inertial forces to remain, as distances from the hazard source grow.

Although dissonance theory has attracted some adherents, more researchers have used locus of control (LOC) as an explanation for hazard perceptions and subsequent behavior.[88] People who believe that their future is determined by powers beyond their control (External LOC) are generally hypothesized as failing to adopt adjustments. Those who believe that personal actions largely determine their lives (Internal LOC) are expected to adopt adjustments. In practice, findings from these studies are equivocal. Some have found strong confirmation,[89] while others have found little or no supporting evidence.[90] The most recent published work combined measures of locus of control with measures of repression-sensitization. This revealed that residents of an earthquake prone community in New Zealand, who took action to protect themselves against the hazard, were more likely to possess Internal LOC and to be moderates who neither repressed nor excessively worried about it.[91] Given the relative paucity of studies which are available, it is not yet possible to judge the importance of LOC as an explanation for hazard perceptions.

Finally, it is necessary to note the existence of a growing literature on limitations affecting human abilities to perceive and judge risk. Slovic and his colleagues have identified and assessed a variety of heuristic devices that influence risk perception. They argue that all humans are affected by such biases as availability of evidence, tendency to believe in personal immunity, and faith in personal ability to control risks. They also underline the persistence of mistaken perceptions of risk. Thus, even when provided with accurate hazard information, populations at risk may not translate this into stimuli for action. Moreover, people may rank and classify hazards in ways that are unrelated to their objective character. Thus, while the degree to which an individual

---

88. A.V. Kirkby, "Perception of Air Pollution as a Hazard and Individual Adjustment to it in Three British Cities," unpublished paper presented to the International Geographical Union, Man and Environment Seminar, Calgary, Alberta, 1972; Baumann and Sims, loc. cit.; Schiff, loc. cit.; Simpson-Housley, op. cit.; John Sims and Duane D. Baumann, "The Tornado Threat: Coping Styles North and South;" Duane D. Baumann and John H. Sims, "Human Response to the Hurricane," in *Natural Hazards: Local, National, Global,* ed. by G.F. White (New York: Oxford University Press, 1974) pp. 25-29.

89. Sims and Baumann, loc. cit.

90. Schiff, loc. cit.

91. Simpson-Housley, op. cit.

feels personally exposed has important influence on risk perception, the degree to which hazards are judged to be unknown or dreaded are also of central importance.[92] If these observations are correct--and there is every reason to believe that they are--it is unlikely that cognitive dissonance and locus of control theories will provide much more than minor partial explanations of hazard perception.

Slovic and other decision scientists have only begun to sketch in the components of a theory of distortion in risk perception. In so doing they have opened up numerous unexplored avenues for further research. Yet ability to estimate risk and perceptual biases are just two of the better researched elements in Kates' and Mileti's models of hazard perception and adjustment. We know very little about other components (e.g. access to information, effectiveness of risk mitigation and adjustments, size of unit of analysis, etc.)

Summary

Hazard Perception Studies are handicapped by lack of a clear cut definition of content and purpose. This blunts cooperative efforts to develop comprehensive research strategies. Most hazards scholars are convinced that what people do about hazards is strongly connected to how they perceive them. Yet, in quantitative terms research throughout the 1970s has been dominated by hazard awareness studies that focused on public understanding of risks, but called for major interpretive leaps in search of explanations of subsequent behavior. Only a minority of scholars probed the processes by which perceptions are linked to decisions.

Nevertheless, investigators have recorded significant progress in: (1) using a diverse array of research techniques; (2) refining major theoretical propositions (i.e. cognitive dissonance theory; locus of control theory); (3) uncovering systematic mechanisms of perceptual bias; (4) disseminating the concept of perception throughout the English speaking community of hazard researchers; and (5) using perception data to begin the design of more effective public information and hazard warning schemes. They must now undertake the more difficult--and less glamorous--task of fleshing

---

92. P. Slovic, B. Fischhoff, and Sarah Lichtenstein, "Behavioral Decision Theory," *Annual Review of Psychology* 28 (1977): 1-27; Paul Slovic, "Toward Understanding and Improving Decisions," in *Human Performance and Productivity*, vol. 2, ed. by W.C. Howell and E.A. Fleshman (Hillsdale, New Jersey: Erlbaum, 1980): 426-450; Paul Slovic and Baruch Fischhoff, "How Safe is Safe Enough? Determinants of Perceived and Acceptable Risk," in *Too Hot to Handle: Social and Policy Issues in the Management of Radioactive Wastes*, ed. by L. Gould and C.A. Walker (New Haven: Yale University Press, 1979); Paul Slovic, Sarah Lichtenstein, and Baruch Fischhoff, "Images of Disaster: Perception and Acceptance of Risks from Nuclear Power," in *Energy Risk Management*, ed. by G. Goodman and W. Rowe (London: Academic Press, 1979) pp. 233-245; P. Slovic, B. Fischhoff, and Sarah Lichtenstein, "Perceived Risk: Psychological Factors and Social Implications," pp. 17-34; Paul Slovic, Baruch Fischhoff, and Sarah Lichtenstein, "Rating the Risks," pp. 14-20; 36-39.

out the existing skeletal understanding of hazard perception principles in the context of hazard adjustment processes. This requires that more attention be paid to the perception and adoption.

CHAPTER 4
TOWARD A THEORETICAL FRAMEWORK FOR LANDSCAPE PERCEPTION

James L. Sell, Jonathan G. Taylor, and Ervin H. Zube

In an earlier study, the authors conducted a survey of landscape perception articles published in some 20 English language journals.[1] The survey covered a 16 year period (1965-1980) of journals from six disciplinary areas: architecture, forestry, outdoor recreation, geography, interdisciplinary-environmental, and behavioral. In general, the journal articles tend to be weighted rather heavily towards field application of landscape quality assessment. Although most application methodologies have an implied theoretical basis, the underlying theory is seldom explicitly articulated in the journal literature. One must turn to other publications--books, symposium proceedings, Forest Service manuals--to investigate the theories behind landscape assessment practice. That is, theory and practice in landscape perception literature tend to be physically segregated.

Appleton[2] comments on the development of a sophisticated array of landscape evaluation techniques within a "theoretical vacuum." In terms of specific landscape perception theory, Appleton appears to be correct; but in terms of theoretical background, landscape perception research is encumbered by the opposite problem: a scattering of diverse theoretical origins. But Appleton's basic point remains salient: that the lack of a unifying theoretical structure does not allow a rational basis for "diagnosis, prescription and prognosis."

Landscape Perception Research Paradigms

In our attempt to put landscape perception literature into an organizational structure, four distinct paradigms of landscape perception research emerged. The first, most obvious dichotomy in

---

1. E.H. Zube, J.L. Sell, and J.G. Taylor, "Landscape Perception: Research, Application and Theory," *Landscape Planning* 9 (1982): 1-33.

2. J. Appleton, "Landscape Evaluation: The Theoretical Vacuum," *Institute of British Geographers, Transactions* 66, (1975): 120-123.

landscape perception literature was according to researcher objectives: pragmatic applications research; or search for meaning in landscape perception (see figure 4.1). The applications research, designed for use by resource managers and planners, was clearly separated according to approach: landscape evaluation conducted by professionals with appropriate training and expertise--the Expert paradigm; or landscape evaluation achieved through empirical testing of public groups--the Psychophysical paradigm. There were also two distinct approaches toward the meaning of landscape perception: the search for the meaning of landscape or the significance of landscape elements to the human mind--the Cognitive paradigm; or search for the meaning of the human-landscape perception interaction--the Experiential paradigm. Each of these research paradigms has evolved from a fairly distinct theoretical base, or combination of bases. These foundations involve the design disciplines and art history, geography, the behavioral sciences--especially psychology, and to some extent the biological sciences and resource management.

The Expert Paradigm: In the expert approach, it is assumed that trained professionals are capable of objectively analyzing landscapes and translating landscape features and characteristics into descriptive assessments of scenic beauty. This paradigm has two theoretical antecedents: the fine arts-design discipline

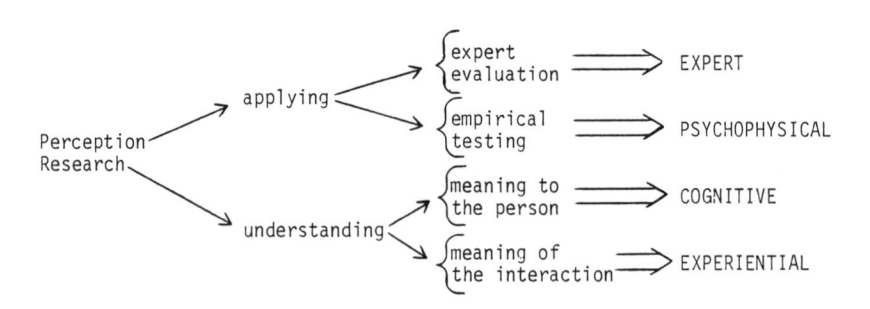

Fig. 4.1 The Emergence of Landscape Perception Research Paradigms

perspective; and the ecological. The fine arts-design tradition can be seen in such works as Litton[3] and Laurie,[4] in which the formal artistic qualities outlined for landscape architecture[5] are translated into assessment criteria for natural beauty. This approach is supported by Carlson[6] who specifically rejects psychophysical empirical assumptions, stating that only expert, non-formalist opinions can be relied upon for landscape assessment. Wright[7] contends that, since aesthetic evaluation is an emotional response within the general public, only trained experts could reliably assess landscape quality.

The ecological tradition is evident in the work of Smardon,[8] and to some degree in that of Leopold,[9] where there is a strong implicit assumption that the natural, unmodified ecosystem carries the greatest aesthetic value. A variation on the expert ecological approach, especially among foresters,[10] assumes that proper resource management will, by definition, yield optimal landscape quality.

The Psychophysical Paradigm: In the psychophysical paradigm, the environment or landscape serves as the source of stimulus to which the individual responds. The theoretical origin for this paradigm stems primarily from stimulus-response assumptions in psychology. It assumes the value of the landscape to be part of its stimulus property, external to the individual, invariant, and perceivable without cognitive processing. This can be seen in Gibson's work[11] on perception, especially the theory of affordances. Applications of stimulus-response assumptions in landscape assessment are evident in the linear regression correlation of Shafer[12] and his colleagues. Daniel and his associates combine

---

3. R.B. Litton, "Aesthetic Dimensions of the Landscape," in *Natural Environments: Studies in Theoretical and Applied Analysis,* ed. J.V. Krutilla (Baltimore: John Hopkins University Press, 1972).

4. I.C. Laurie, "Aesthetic Factors in Visual Evaluation," in *Landscape Assessment: Values, Perceptions, and Resources,* ed. E.H. Zube, R.O. Brush, and J.G. Fabos (Stroudsburg, PA: Dowden, Hutchinson, and Ross, 1975).

5. See, for example: J.O. Simonds, *Landscape Architecture: The Shaping of Man's Natural Environment* (New York: F.W. Dodges, 1961); H.V. Hubbard and T. Kimball, *An Introduction to the Study of Landscape Design* (New York: Macmillan, 1977).

6. A.A. Carlson, "On the Possibility of Quantifying Scenic Beauty," *Landscape and Planning* 4 (1977): 131-172.

7. G. Wright, "Appraisal of Visual Landscape Qualities in a Region Selected for Accelerated Growth," *Landscape Planning* 1 (1974): 307-327.

8. R.C. Smardon, "Assessing Visual-Cultural Values of Inland Wetlands in Massachusetts," in *Landscape Assessment: Values, Perceptions, and Resources,* ed. E.H. Zube, R.O. Brush and J.G. Fabos (Stroudsburg, PA: Dowden, Hutchinson, and Ross, 1975).

9. L. Leopold, "Landscape Aesthetics," *Natural History* 73, no. 4 (1969), pp. 36-45.

10. L.L. Streeby, "Scenic Management Impact on Other Forest Activities," *Journal of Forestry* 68 (1970): 430-432.

11. J.J. Gibson, "The Theory of Affordances," in *Perceiving, Acting and Knowing: Toward an Ecological Psychology,* ed. R. Shaw and J. Bransford (Hillsdale, N.J.: Erlbaum, 1977).

12. E.L. Shafer, "Perception of Natural Environments," *Environment and Behavior* 1 (1969): 71-82.

stimulus-response with methodologies developed in signal detection research[13] for scenic assessment of forests, roads and air quality. Buhyoff and his colleagues have applied scenic quality scaling techniques extensively to aesthetic impacts of Southern Pine Beetle damage.[14] Zube and others have applied Q-sort and numerical rating techniques to landscape evaluations.[15] Much of the recreational landscape perception literature falls into the psychophysical paradigm[16] as might be expected of a discipline with such heavy reliance on assessment of user preferences.

The Cognitive Paradigm: The cognitive approach deals with landscape quality as a construction built up in the human mind, usually from visual modes of information gathering. This paradigm supports several research approaches stemming from several theoretical origins. Psychobiological approaches derive primarily from Berlyne's arousal theory in which aesthetic preferences correspond to biological importances.[17] Helson, in turn, developed adaptation level theory, and Wohlwill[18] follows both of these traditions with the perception theory of optimal levels of arousal. Another psychobiological approach is that of sentics,[19] which postulates characteristic, recognizable shapes of neural patterns of emotion, which parallel landscape features. Several evolutionary-adaptational theories have been applied to landscape perception. Appleton[20] draws upon a broad base of theoretical material, including art-history, philosophy and ethology to propose that human evolution as hunting (and hunted) creatures has provided

---

13. T.C. Daniel and R.S. Boster, *Measuring Landscape Esthetics: The Scenic Beauty Estimation Method* (1976), USDA Forest Service Research Paper RM-167.

14. G.J. Buhyoff, W.A. Leuschner, and J.D. Wellman, "Aesthetic Impacts of Southern Pine Beetle Damage," *Journal of Environmental Management* 8 (1979): 261-267.

15. E.H. Zube, "Rating Every Day Rural Landscapes of the Northeastern U.S.," *Landscape Architecture* 63 (1973): 92-97.

16. See, for example: G.L. Peterson, "Evaluating the Quality of the Wilderness Environment," *Environment and Behavior* 6 (1974): 169-193; T.A. Heberlein and P. Dunwiddie, "Systematic Observation of Use Levels, Campsite Selection and Visitor Characteristics at a High Mountain Lake," *Journal of Leisure Research* 11, no. 4 (1979), pp. 306-316; R.L. Levine and E.E. Langenau, Jr., "Attitudes Toward Clearcutting and Their Relationships to the Patterning and Diversity of Forest Recreation Activities," *Forest Science* 25 (1979): 317-327.

17. D.E. Berlyne, *Aesthetics and Psychobiology* (New York: Appleton, Century-Crofts, 1971); *Conflict, Arousal, and Curiosity* (New York: McGraw-Hill, 1960).

18. J.F. Wohlwill, "Environmental Aesthetics: The Environment as Source of Affect," in *Human Behavior and Environment* Vol. 1, ed. I. Altman and J.F. Wohlwill (New York: Plenum Press, 1976).

19. See, for example: B.B. Greenbie, "Problems of Scale and Context in Assessing a Generalized Landscape for Particular Persons," in *Landscape Assessment: Values, Perceptions, and Resources,* ed. E.H. Zube, R.O. Brush, and J.G. Fabos (Stroudsburg, PA: Dowden, Hutchinson and Ross, 1975); M. Clynes, "Toward a Theory of Man: Precision of Essentic Form of Living Communications," in *Information Processing in the Nervous System,* ed. N. Leibovic and J.C. Eccles (New York: Springer-Verlag, 1969).

us with a preference for prospects (where one can see) and refuges (where one cannot be seen). Charlesworth's theory that species recognize and prefer environments to which they are best adapted has been carried forward into landscape perception research by the Kaplans[21] in terms of information processing theory. Landscape preferences are felt to be related to the adaptive need to make sense of and be stimulated by the environment.

The role of group values[22] and of personality type[23] as they affect landscape perception have been explored. Some correlation between personality and environmental response has been suggested, but to date no general theory has been put forward. Several contributions within the cognitive landscape-perception paradigm deal with sociocultural effects on landscape evaluation, including the Buhyoff et al.[24] study of landscape architects' ability to adopt client perspectives; Riley's[25] consideration of the influence of special interest groups in shaping landscape; Sonnenfeld's[26] comparison of Alaskan and Delaware residents' perceptions; and Zube's[27] comparisons among various professional and general public groups, as well as cross-cultural comparisons.[28] It is within the cognitive paradigm where the majority of the natural vs. man-made landscape comparisons have been made.[29]

---

20. J. Appleton, *The Experience of Landscape* (New York: John Wiley and Sons, 1975).

21. See R. Kaplan, "The Green Experience," in *Humanscape,* ed. S. Kaplan and R. Kaplan (North Scituate, MA: Duxbury, 1979); S. Kaplan, "Perception and Landscape: Conceptions and Misconceptions," in *Our National Landscape,* ed. G.H. Elsner and R.C. Smardon (USDA Forest Service, General Technical Report PSW-35, 1979); S. Kaplan, "An Informal Model for the Prediction of Preference," in *Landscape Assessment: Values, Perceptions, and Resources,* ed. E.H. Zube, R.O. Brush, and J.G. Fabos (Stroudsburg, PA: Dowden, Hutchinson, and Ross, 1975).

22. E.C. Penning-Rowsell, "The Social Value of English Landscapes," in *Our National Landscape,* ed. G.H. Elsner and R.C. Smardon (USDA Forest Service, General Technical Report PSW-35, 1979).

23. See K.G. Craik, "Individual Variations in Landscape Description," in *Landscape Assessment: Values, Perceptions, and Resources,* ed. E.H. Zube, R.O. Brush, and J.G. Fabos (Stroudsburg, PA: Dowden, Hutchinson, and Ross, 1975); R.R. Little, "Specialization and the Varieties of Environmental Experience: Empirical Studies Within the Personality Paradigm," in *Experiencing the Environment,* ed. S. Wapner, S.B. Cohen, and B. Kaplan (New York: Plenum Press, 1975).

24. G.J. Buhyoff, J.D. Wellman, H. Harvey, and R.A. Fraser, "Landscape Architects' Interpretations of People's Landscape Preferences," *Journal of Environmental Management* 6 (1978): 255-262.

25. R.B. Riley, "Speculations on the New American Landscapes," *Landscape* 24, no. 3 (1980), pp. 1-9.

26. J. Sonnenfeld, "Equivalence and Distortion of the Perceptual Environment," *Environment and Behavior* 1 (1969): 83-99.

27. E.H. Zube, "Rating Every Day Rural Landscapes of the Northeastern U.S."

28. E.H. Zube and D.G. Pitt, "Cross-Cultural Perceptions of Scenic and Heritage Landscapes," *Landscape Planning* 8 (1981): 69-87.

29. See R.S. Ulrich, "Visual Landscapes and Psychological Well-Being," *Landscape Research* 4, no. 1 (1979), pp. 17-23; J.F. Wohlwill and G. Harris, "Response to Congruity or Contrast for Man-Made Features in Natural Recreation Settings," *Leisure Science* 3 (1980): 349-365; J.F. Wohlwill and H. Heft, "A Comparative Study of User Attitudes Towards Development and Facilities in Two Contrasting Natural Recreation Areas," *Journal of Leisure Research* 9, no. 4 (1971), pp. 264-280; C.A. Acking and G.J. Sorte, "How do We Verbalize What We See?," *Landscape Architecture* 63 (1973): 120-125.

The Experiential Paradigm: This final research orientation centers on the experience or phenomenon of human-environment interaction. In terms of theoretical antecedents, this approach relies in part upon art-history and literary traditions. The emphasis on experience suggests that aesthetic quality lies both in the landscape and in the meaning of the landscape to people; indeed they are difficult to separate from the context of the particular situation, and from other emotional experiences. Within this context it is difficult to employ techniques for analysis in other than unstructured phenomenological exploration[30] or reviews of literature and art.[31] Focus, within experiential landscape perception research, has been on ordinary landscapes;[32] sense of place;[33] home;[34] historical and cultural expressions;[35] and on visual blight.[36]

An Alternative Structuring of Landscape Perception Research

This paradigm structuring of landscape perception literature is corroborated by an independent analysis conducted by other perception researchers. Daniel and Vining[37] have categorized approaches to landscape perception research into five conceptual models which closely parallel our four paradigms. Their conceptual models are: Ecological, Formal Aesthetic, Psychophysical, Psychological, and Phenomenological. Their Ecological and Formal Aesthetic are analogous to our Expert, and the Psychological and Phenomenological are analogous to our Cognitive and Experiential. Psychophysical is common to both research structures. Daniel and Vining evaluate each of their conceptual models in reference to

---

30. See D. Seamon, *A Geography of the Lifeworld* (New York: St. Martin's, 1979); M. Merleau-Ponty, *Phenomenology of Perception,* translated by C. Smith, (New York: Humanities Press, 1962).

31. See D. Lowenthal, "Finding Valued Landscapes," *Progress in Human Geography* 2 (1978): 373-418; D.E. Sopher, "The Landscape of Home: Myth, Experience, Social Meaning," in *The Interpretation of Ordinary Landscapes,* ed. D.W. Meinig (Oxford: Oxford University Press, 1979).

32. D.W. Meinig, ed., *The Interpretation of Ordinary Landscapes* (Oxford: Oxford University Press, 1979).

33. Y.F. Tuan, *Space and Place: The Perspective of Experience* (Minneapolis: University of Minnesota Press, 1977).

34. D.E. Sopher, "The Landscape of Home: Myth, Experience, Social Meaning."

35. See D. Lowenthal, "Age and Artifact: Dilemmas of Appreciation," in *The Interpretation of Ordinary Landscapes,* ed. D.W. Meinig (Oxford: Oxford University Press, 1979); D. Lowenthal and H.C. Prince, "English Landscape Taste's," *The Geographical Review* 55 (1965): 186-222.

36. See P.F. Lewis, D. Lowenthal, and Y.F. Tuan, *Visual Blight in America,* Commission of College Geography Resource Paper no. 23 (Washington, D.C.: Association of American Geographers, 1973); J.B. Jackson, "To Pity the Plumage and Forget the Dying Bird," in *Landscapes: Selected Writing of J.B. Jackson,* ed. E.H. Zube (Amherst, MA: The University of Massachusetts Press, 1970).

37. T.C. Daniel and J. Vining, "Methodological Issues in the Assessment of Visual Landscape Quality," in *Behavior and the Natural Environment* 6, ed. I. Altman and J. Wohlwill (New York: Plenum Press, 1983).

reliability, sensitivity, validity and utility. Reliability is the consistency of findings from repeated measurements; sensitivity is the ability to measure actual differences; validity is the congruence between what is being measured and what is purported to be measured; and utility is whether the findings can be used for what they were intended. Table 4.1 provides a summary evaluation of paradigms based on the Daniel and Vining evaluative criteria.

Daniel and Vining suggest the reliability of expert and experiential paradigms is questionable because of idiosyncratic approaches and because of mixed results to date in the few studies of agreement among independent experts. In addition, they conclude that the validity of the expert paradigm is "still open to question as is the experiential." However, the latter may provide a more valid assessment of landscape experience than other models or paradigms because of its greater attention to human oriented concerns and less attention to details of the landscape. Psychophysical and cognitive paradigms have, by virtue of the methods employed, been subjected to the most extensive tests for reliability and validity.

Daniel and Vining approach the question of utility from several perspectives including ease of application to planning and management problems, sensitivity to variations among landscapes, and the reliability and validity of methods. As suggested above, reliability and validity of the expert and experiential paradigms are in question. The expert does, however, meet the criterion for ease of application in that methods have been specifically developed with that objective in mind. Both the cognitive and the experiential paradigms are found wanting with reference to ease of application to planning and management problems, as the landscape is not addressed in physical terms that are useful or manipulatable by planning and management procedures.

Sensitivity of the expert paradigm is uncertain and varies with the way in which landscapes are categorized, both ecologically and visually. Methods tend to group landscapes ordinally into high, medium and low quality classes, thus masking distinctions among some landscapes. Daniel and Vining assert that the methods employed in the psychophysical and cognitive paradigms can yield ordinal or interval scale data of high sensitivity. The most sensitive paradigm, however, is the experiential. Yet, this high level of sensitivity mitigates against its utility as each person in each situation is viewed as a unique event.

TABLE 4.1

EVALUATION OF PARADIGMS

Evaluative* Criteria	Paradigms			
	Expert	Psychophysical	Cognitive	Experiential
Reliability	Low	High	High	Low
Sensitivity	Low	Mod	Mod	Mod
Validity	Low	High	High	Mod
Utility	Mod**	Mod	Low	Low

*After Daniel and Vining (1983) **Mod = moderate.

These examinations of the literature lead to the primary conclusion that landscape perception research efforts are marked by separation: separation by paradigm; separation by theoretical origin; and, as indicated at the outset, segregation of theory into books or symposium proceedings and of applications presented in the journal literature. A further form of separation has been the tendency in much research to focus on elements--elements of the human and/or components of the landscape--rather than to focus on the perceptual interaction. This elemental focus leads to descriptive research, concentrating on the "what" of landscape perception rather than on the "how" or "why." Thus, at this time when perception research stems from various theoretical origins and part has grown out of a pragmatic need generated by legal mandates for management of landscape quality, the most pressing need is for a basic model to which landscape perception research and theory can be fit into a whole.

A Model of Landscape Perception

An interaction model is constructed to provide a framework for examination and comparison of landscape perception research and the four different paradigms. In this model, landscape perception is considered a function of the interaction of humans and landscapes[38] and the outcomes of the perceptual interaction feed back to modify the human, landscape, and interaction components (see figure 4.2). Through this model, the different conceptual perspectives and assumed relevant components of the human, landscape and outcome--brought by the various research paradigms--can be compared

---

38. E.H. Zube, R.O. Brush, and J.G. Fabos, *Landscape Assessment: Values, Perceptions and Resources* (Stroudsburg, PA: Dowden, Hutchinson, and Ross, 1975); W.H. Ittelson and H. Cantril, *Perception: A Transactional Approach* (New York: Doubleday, 1954).

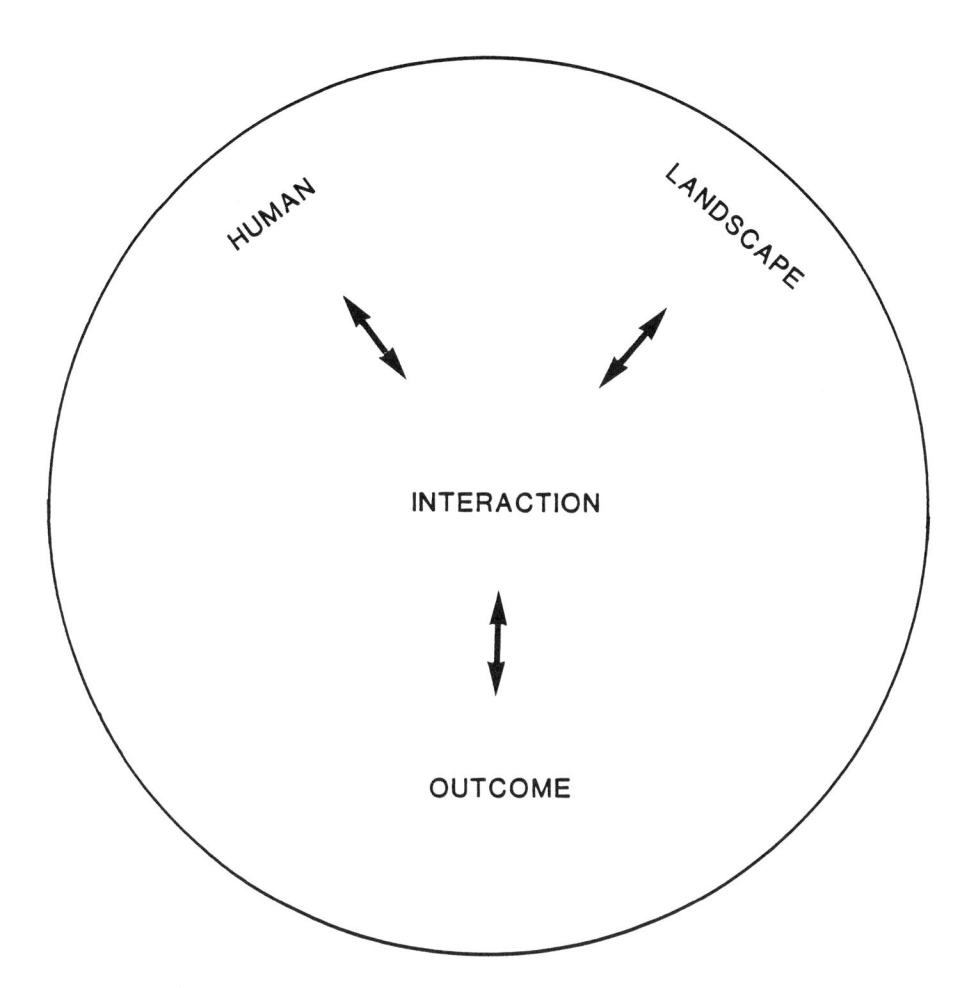

Figure 4.2 Landscape Perception Interaction Process

and integrated.

Concepts of the Human: The human concept refers to the implicit or explicit assumptions about the nature of humanity--whether the observer is highly skilled; acting as a respondent, an information processor, or an active participant in the interaction process--and those features of human beings that are appealed to in a particular interaction with landscape. Components of the human encompass past experience, knowledge, expectations, personality, and the sociocultural context of individuals and groups.

Landscape Properties: The landscape component includes both individual elements of landscapes and whole landscapes as entities. These refer to those tangible or intangible elements or relationships in the landscape that contribute significantly to the perceptual interaction process. Landscape concepts may assume salience based on principles of art, design, or ecology; on properties that are manipulatable through management; on the association with obtaining information and meaning; or on the world of everyday experience. Tangible elements might be "downed wood" or "water edge;" intangible elements might include "naturalism," "composition," or "mystery."

Interaction: The perceptual interaction itself encompasses human-landscape, group-landscape, human-human, and human-group permutations. Concepts of the interaction include active or passive, purposeful or accidental, unique or habitual experiences.

Interaction Outcomes: Perceptual interaction between humans and landscapes result in outcomes which in turn affect both the human and landscape components and their subsequent inputs into the interaction. Outcomes may be either tangible, such as a state of physical change; or intangible, such as a state of mind or feeling. Implicit or explicit assumptions about the nature of perceptual interactions emphasize different interaction outcomes as important: statements of landscape quality, numerical expressions of perceived values, ratings of preference, meaning, adaptation, habitual behavior, statements of landscape taste, or enhancement of sense-of-self.

The relationships between the model elements--human, landscape, and outcome--can be used to organize the various approaches into one system of landscape perception. In this schema, it appears that the expert and psychophysical paradigms focus mainly on the landscape, the cognitive emphasizes the human side, and the experiential is most concerned with the process of the interaction and the outcome. While it is yet to be proven that all studies in all paradigms can be made to mesh into one overriding system, it is our contention that each offers important insights into the development of a single theoretical approach with practical utility. Furthermore, such an overall schema can be seen as a natural outgrowth of one of the most significant psychological approaches to perception of environment research, that of William H. Ittelson.

Transactionalism and Landscape Value

The conceptual model presented here as a theoretical framework for landscape perception is a transactional one in which human and landscape values grow out of a situation of mutual influence. Transactionalism as a psychological approach was adapted by Ames and his students[39] from the work of John Dewey[40] that emphasizes the importance of the experience in defining the identity of the entities in transactions. The transactional approach to environmental perception has been best articulated by Ittelson.[41] At the core of Ittelson's conception of perception is the distinction between environments and objects, such as are usually studied in laboratory perception research:

> Objects require subjects--a truism whether one is concerned with the philosophical unity of the subject-object duo, or is thinking more naively of the object as a 'thing' which becomes a matter of psychological study only when observed by a subject. In contrast, one cannot be a subject of an environment, one can only be a participant. The very distinction between self and nonself breaks down; the environment surrounds, enfolds, engulfs, and no thing, and no one can be isolated and identified as standing outside of and apart from it.[42]

Ittelson provided further guidance for the structuring of a theoretical framework by identifying a set of minimum considerations "which must be taken into account in any adequate study of environment perception."[43] These considerations appear below as we have adapted them to apply to landscape perception:

1. Landscapes surround. They permit movement and exploration of the situation and force the observer to become a participant.
2. Landscapes are multimodal. They provide information that is received through multiple senses and that is processed simultaneously.
3. Landscapes provide peripheral as well as central information. Information is received from behind the participant as well as from in front, from outside the focus of attention as well as within.
4. Landscapes provide more information than can be used. They can simultaneously provide redundant, inadequate, ambiguous, conflicting and contradictory information.
5. Landscape perception always involves action. Landscapes cannot be passively observed; they provide opportunities for action, control and manipulation.
6. Landscapes call forth actions. They provide symbolic meanings and motivational messages that can call forth purposeful actions.

---

39. See Ittelson and Cantril, op. cit.

40. See J. Dewey and A.F. Bentley, *Knowing and The Known* (Boston, MA: Beacon, 1949).

41. W.H. Ittelson, "Environmental Perception and Contemporary Perceptual Theory," in *Environment and Cognition,* ed. W.H. Ittelson (New York: Seminar Press, 1973).

42. Ibid. pp. 12-13.

43. Ibid. pp. 12-16.

7. Landscapes always have ambiance. They are almost always encountered as part of a social activity, they have a definite aesthetic quality and they have a systemic quality (various components and events are related).

It is Ittelson's contention that environments, or landscapes, cannot be studied in the same way as objects in the laboratory. The first four conditions in the preceeding list are stimulus properties of environments which objects "cannot or usually do not possess." The last three conditions, concerned with actions, meanings, motivations, and ambiance, are essential additions for landscape research, as they relate to properties of the transaction. All of the conditions are based on the recognition that the central issue is the transaction between humans and landscape. These points made by Ittelson were also examined by Unwin,[44] who stressed the qualities of surrounding, peripheral information, active perceptual mode, and meaning or atmosphere as most important for landscape evaluation. It is useful at this point to discuss the contributions of the four landscape paradigms in terms of Ittelson's seven properties. To aid in this discussion, table 4.2 provides a generalized summary of these contributions.

TABLE 4.2

PARADIGM CONTRIBUTIONS TO A TRANSACTIONAL
APPROACH TO PRECEPTION OF ENVIRONMENT

Perceptual* Properties	Paradigms			
	Expert	Psychophysical	Cognitive	Experiential
Surrounding	Some	Little	Some	Considerable
Multimodality	Little	None	None	Little
Peripherial Information	Some	Little	Little	Little
Excessive Information	Little	Little	Considerable	Little
Action	Some	Little	Some	Considerable
Action Symbolism	Little	Little	Considerable	Considerable
Ambiance	Little	Little	Some	Some

*After Ittelson (1973)

---

44. K.I. Unwin, "The Relationship of Observer and Landscape in Landscape Evaluation," *Institute of British Geographers, Transactions* 66 (1975): 130-134.

1. *Landscapes surround*. There are two major aspects to this notion: that landscapes are experienced subjectively and their perception is a process requiring action. Unlike an object or a picture, a landscape is viewed from inside, not as something detached from the viewer which can be examined from the outside, i.e., "objectively." Another quality of surrounding is that landscapes "are necessarily larger than that which they surround." In order to perceive a landscape, one must explore it--move through it, manipulate it, participate in the activities of the setting. The latter two kinds of exploration will be discussed in later sections, of immediate concern here is the quality of movement.

Although the art expert approach tends to consider the landscape as an object viewed in the same way as a painting on canvas, there are some proponents of the idea that aesthetic qualities must be seen in the field rather than through such surrogates as pictures.[45] There is also an awareness that the observer moves through the landscape to change his perspective.[46] Litton,[47] for example, asserted that the changes in observer distance from a scene, vertical position (above, below, or on the same level as the scene), and in the sequence of successive views, must all be included in scenic planning.

The ecological expert view tries to place people inside the landscape, although the effect of their participation is often thought of as less than benign.[48] There is also a tendency to dichotomize between natural and human systems rather than seeing them as one landscape; this dichotomy is often overlain by broad value judgments, as can be seen in the following statement:

> Natural systems are balanced, intricate, and dynamic. Human society, on the other hand, is internally divided, competitive, wasteful, and unstable.[49]

Many of the researchers in the psychophysical paradigm ignore the major difference between landscape as external stimulus object

---

45. See P. Dearden, "A Statistical Method for the Assessment of Visual Landscape Quality for Land-Use Planning Purposes," *Journal of Environmental Management* 10 (1980): 51-68; Laurie, op. cit.

46. D.B. Carruth, "Assessing Scenic Quality: Transmission Line Siting," *Landscape* 22, no. 1 (1977), pp. 31-34; R.E. Burke, "National Forest Visual Management: A Blend of Landscape and Timber Management," *J. Forestry* 73, no. 2 (1975), pp. 767-770; R.B. Litton, *Forest Landscape Description and Inventories*, USDA Forest Service Research Paper PSW-49 (Berkeley, California: Pacific Southwest Forest and Range Experiment Station, 1968).

47. Litton, 1968, op. cit.

48. R.J. Reimold, M.A. Hardisky, and J.H. Phillips, "Wetland Values--A Non-Consumptive Perspective," *Journal of Environmental Management* 11 (1980): 77-85; J.D. Ovington, K.W. Groves, P.R. Stevens, and M.T. Tanton, "Changing Scenic Values and Tourist Carrying Capacity of National Parks, An Australian Example," *Landscape Planning* 1 (1974): 35-50; Smardon, op. cit.

49. Reimold et al., op. cit., p. 78.

and as surrounding environment, thus providing a rationale for the use of pictures as surrogates. A few studies have attempted to document the parallels between pictures and landscapes, rather than emphasize the major differences.[50] Others have attempted to collect on-site evaluations.[51]

The cognitive approach is primarily concerned with the subjective meaning of landscape, thus environments have this surrounding quality but also are somewhat detached from what is seen as most important--the mental processing of meaning. This mental processing basis appears to allow the use of surrogates for landscapes in their research, because the stimulus for study appears to result in similar aesthetic responses. For example, S. Kaplan[52] stresses the importance of making sense of the environment and feeling involved with it as underlying aesthetic responses to both two dimensional visual arrays and actual landscapes. Sonnenfeld[53] goes somewhat further, stressing an insider-outsider difference in aesthetic response to places that is related to adaptational needs.

This insider-outsider difference is also seen in the experiential paradigm.[54] This quality of surrounding is related to the inseparability of the experienced and the experiencer--

> Landscape . . . is a continuous and enveloping experience. There are no frames to a view. The vista changes as we move through and in a landscape.[55]

Indeed, the continuity between person and landscape is important. Relph[56] warns against "visual alienation," a tendency to over-objectify, in the viewing of landscapes.

2. *Landscapes are multimodal.* In all paradigms there is a tendency to concentrate on the visual mode of experience; little has been done to examine how other senses contribute on their own or work together in the perception of a landscape. There are a few in the expert and experiential paradigms who include non-visual modes

---

50. Daniel and Boster, op. cit.; S. Shuttleworth, "The Use of Photographs as an Environment Presentation Medium in Landscape Studies," *Journal of Environmental Management* 11 (1980): 61-76.

51. R.J. Foster and E.L. Jackson, "Factors Associated with Camping Satisfaction in Alberta Provincial Park Campgrounds," *Journal of Leisure Research* 11, no. 4 (1975), pp. 292-306; G.J. Buhyoff, "A Methodological Note on the Reliability of Observationally Gathered Time-Spent Data," *Journal of Leisure Research* 11, no. 4 (1979), pp. 334-342; Heberlein and Dunwiddie, op. cit.

52. S. Kaplan (1979), op. cit.

53. Sonnenfeld, op. cit.

54. Seamon, op. cit.; Y.F. Tuan, "Visual Blight: Exercises in Interpretation," in *Visual Blight in America,* ed. P.F. Lewis, D. Lowenthal, and Y.F. Tuan (Washington, D.C.: Association of American Geographers, 1973).

55. P.T. Newby, "Towards an Understanding of Landscape Quality," *Landscape Research* 4, no. 2 (1979), p. 13.

56. E.C. Relph, "To See With the Soul of the Eye," *Landscape* 23 (1979): 28-34.

of sensing.[57] The following commentary suggests one approach to the study of the total sensual experience:

> An aesthetic appreciation of marshes, swamps, and other wetlands is essentially a sensual one. Wetlands stimulate the vision, hearing, sense of smell, touch, and taste in ways that have been recorded by painters, musicians and writers of many ages.[58]

3. *Landscapes provide peripheral as well as central information.* Because the individual is surrounded by information, some mechanism must be used to determine what should be the central focus of attention. Most researchers in the cognitive and psychophysical paradigms, especially, tend to define the focus of attention themselves by presenting slides or asking about certain scenes in the field. Daniel and Boster[59] have attempted to get around this experimenter-determined focus of attention by randomizing the direction of pictures taken for later presentation to subjects. The experiential approach tends to leave determination of attention to the individual and encourage an overall sensitivity.[60] The "experts," especially those using the artistic-design approach, have looked at focal points of attention as part of the formal compositional quality of the landscape. Burke,[61] for example, discusses the qualities of "axis," "convergence," "co-dominance," and "enframement" as stimuli for drawing the observer's attention.

4. *Landscapes provide more information than can be used.* No one uses all the information available in a landscape, there is too much to process and much that is not relevant to the observer's purpose. Awareness of the complexity of information in a landscape is present in all four paradigms. The best treatment of complexity as an aesthetic quality is most probably found in the cognitive paradigm. In the psychobiological viewpoint of Berlyne[62] and Wohlwill,[63] the relationship between complexity and affect is an inverted U-shaped function, too little or too much information is unpleasant while a moderately stimulating yet not overwhelming amount is pleasant. Acking and Sorte[64] attempted to establish a link between complexity and practical design problems by relating it

---

57. Reimold et al., op. cit.; Relph, op. cit.

58. Reimold et al., op. cit., p. 79.

59. Daniel and Boster, op. cit.

60. S.R. Aiken, "Towards Landscape Sensibility," *Landscape* 20, no. 3 (1976), pp. 20-28.

61. Burke, op. cit.

62. Berlyne, op. cit.

63. Wohlwill, op. cit.

64. Acking and Sorte, op. cit.

to the number of elements in a scene. To S. Kaplan,[65] the complexity of a scene is important in establishing the degree of involvement--the more there is "going on," the more interested the observer will be. Complexity also must be balanced by "coherence," the degree to which the scene can be organized. To Kaplan, coherence is best where the view can be organized into a few dimensions:

> People can only hold a certain amount of information in what is called their "working memory" at one time . . . Thus, rather than being able to remember a certain number of individual details or facts, people seem to be able to hold on to a few distinct larger groupings of information. The current evidence suggests that most people are able to hold approximately five such chunks or units in their working memory at once.[66]

5. *Landscape perception always involves action.* The perception of landscape, even as "scenery" involves some sort of action, because it encompasses the human and the context of the experience is tied in with this action. Researchers most willing to incorporate the role of action in the experience of landscape are those in the experiential paradigm. Of special interest to experiential researchers is the landscape as it is perceived in some sort of creative activity.[67] Matro, for instance, suggests that examination of literary sources will provide useful information:

> . . . a good deal of modern literature provides a way to focus on perception in action, a way of looking at what happens when an individual sensibility encounters its influences, its environment, and in the process realizes its dependence on these factors in the very act of trying to express or interpret them.[68]

Those concerned with recreation resource management have also commented on differences in landscape preferences and attitudes between different types of activity patterns. Among psychophysical researchers, Levine and Langenau[69] provided evidence that attitudes toward clearcutting are related to differences in recreation activities, and Knopp, et al.[70] found river users have different types of "preference packages." Experts in silviculture have long been aware of the opportunities for manipulation of scenic forest qualities, and have tended to assert that the same actions that enable a sustained timber harvest, such as thinning and selective

---

65. S. Kaplan (1979), op. cit.

66. Ibid., p. 244.

67. See A. Kobayashi, "Landscape and the Poetic Act: The Role of Haiku Clubs for the Issei," *Landscape* 24, no. 1 (1980), pp. 42-47; Relph, op. cit.; Lowenthal, op. cit.; T. Matro, "Poetry and the Effort of Perception," *Landscape* 22, no. 2 (1978), pp. 14-18; J. Zaring, "The Romantic Face of Wales," *Association of American Geographers, Annals* 67 (1977): 397-418.

68. Matro, op. cit., p. 17.

69. Levine and Langenau, op. cit.

70. T.B. Knopp, G. Ballman, and L.G. Merriam, Jr., "Toward a More Direct Measure of River User Preferences," *Journal Leisure Research* 11, no. 4 (1979), pp. 317-326.

cutting, will also improve forest aesthetics.71 Perlman72 sees a functional difference between rural and city dwellers in their appreciation of landscape. Because city residents use the rural landscape for leisure activity, they tend to possess an "extreme sensitivity to anything that is perceived to damage the countryside." This sensitivity tends to conflict with the wishes of rural dwellers to improve their lot in life through such things as modernization of agriculture. The notion of control was addressed by a researcher whose approach is primarily in the cognitive paradigm.73 In Riley's discussion of the relationship between landscape attitudes, activity and technology, special stress was placed on the feeling of control:

> Our feeling of control over the environment and our sense of competence can be as important as the attributes of the environment itself.74

Perhaps the best way to round out this section is to note that an active view of landscape perception must incorporate more than the aesthetics of looking at scenery. Many different types of activities add many different qualities to the human condition, and must be subsumed under a larger set of values than simply aesthetic. In evaluating wetlands, for example;

> We can sift through an infinite variety of human experiences in wetlands to enumerate non-consumptive uses. Some of these might include sailing or boating through the marshes, tasting and smelling salt spray, being engulfed by fog while hearing fog horns or bell buoys, squeezing marsh mud between one's toes, seining for fish or harvesting shellfish.75

6. *Landscapes call forth actions*. In addition to being experienced actively, landscapes possess a great variety of possibilities. Awareness of these possibilities is what Ittelson refers to as "motivational messages" in landscape perception. But it is the meaning of these messages, the symbolism of landscape, that is most important in affecting the directions taken by the action.76

Symbolism is most prominent in the cognitive and experiential paradigms. Appleton77 believes the long evolution of humans as hunters has given them an underlying aesthetic reaction to places

---

71. A.A. MacDonald, "Looking Forward. Timber Forests Versus Amenity Forests," *Forestry* supplement; *Fifty Years On* (1969): 122-124; J.D. Matthews, "Forestry and the Landscape," *Forestry* 40 (1967): 15-20.

72. R. Perelman, "Contribution to Reflections on Rural Landscapes," *Landscape Planning* 7 (1980): 223-228.

73. Riley, op. cit.

74. Ibid., p. 9.

75. Reimold, op. cit., p. 79.

76. Ittelson, op. cit.

77. Appleton, op. cit.

that provide advantages in hunting, that is, places to hide and from which large areas can be surveyed. Others[78] consider the symbolism of implied human influence as very important in landscape aesthetics. Riley[79] sees another type of meaning in the American landscape--as a product to be consumed, "as a physical resource for specialized interests centered on recreation and merchandizing." This meaning affects both the way the landscape is developed and American attitudes toward it.

Discussions of the meaning of landscape to Americans are also found in the experiential paradigm,[80] as are commentaries on English landscape tastes[81] and romantic views of Wales.[82] Others[83] assert the importance of individual experience in landscape symbolism;

> It is experience that allows us to attribute meanings. If we have experience of a landscape then that landscape is meaningful for us.[84]

7. *Landscapes always have an ambiance*. This "atmosphere" surrounding a landscape is, to Ittelson, difficult to define but overriding in importance. He does specify, however, three features contributing to ambiance. First is the social dimension; landscapes are only rarely encountered by lone individuals, other people are usually present and the encounter is part of a social activity. Next, landscapes always have an aesthetic, an affective property--"esthetically neutral objects can be designed, esthetically neutral environments are unthinkable."[85] The third feature associated with this ambiance is a "systemic quality:"

> The various components and events relate to each other in particular ways which, perhaps more than anything else, serve to characterize and define the particular environment.[86]

The notion of a social, and beyond that, a cultural context to landscape encounters has not been well explored. Several psychophysical studies[87] included crowding as a factor in

---

78. R.W. Hodgson and R.L. Thayer, "Implied Human Influence Reduces Landscape Beauty," *Landscape Planning* 7 (1980): 171-179; J.F. Wohlwill, "What Belongs Where: Research on the Fittingness of Man-Made Structures in Natural Settings," *Landscape Research* 3, no. 3 (1978), pp. 3-5.

79. Riley, op. cit.

80. D. Lowenthal, "The Bicentennial Landscape: A Mirror Held Up to the Past," *The Geographical Review* 67 (1977): 249-267; D. Lowenthal, "The American Scene," *The Geographical Review* 58 (1968): 61-88; E.H. Zube, ed., *Landscapes: Selected Writings of J.B. Jackson* (Amherst: University of Massachusetts Press, 1970).

81. Lowenthal and Prince, 1965, op. cit.

82. Zaring, op. cit.

83. See Newby, op. cit.; Aiken, op. cit.

84. Newby, op. cit., p. 16.

85. Ittelson, op. cit.

86. Ibid., p. 15.

87. G.O. Ewing and T. Kulka, "Revealed and Stated Preference Analysis of Ski Resort Attractiveness," *Leisure Sciences* 2 (1979): 249-275; E.G. Carls, "The Effects of People and Man-Induced Conditions on Preference for Outdoor Recreation Landscapes," *Journal of Leisure Research* 6 (1974): 113-124.

attractiveness of recreation landscapes. In the cognitive approach, researchers have examined differences between social groups and among cultures[88] and stressed the importance of cultural attitudes.[89] Among the experiential group, Wilkinson[90] found that loneliness and the lack of other women were important in the way women perceived the Oregon Trail, and Lowenthal[91] has discussed the importance of the cultural context.

Lowenthal[92] has also stressed the importance of relationships with the past in the perception of landscapes. Indeed, the recognition that there is some sort of systemic relationship between individual experience and landscape components is at the core of the experiential perspective.[93] Relationships among landscape components have been investigated by cognitive researchers. S. Kaplan,[94] for example, considers coherence and legibility to be significant factors in preference judgements. A similar view is found expressed by Wohlwill,[95] in which the quality of unity is seen as important.

The last quality of ambiance set forth by Ittelson is an "aesthetic" one. It is Ittelson's belief that apart from aesthetic judgements of a cognitive nature, there is also a prejudgemental affective system that sets the tone for subsequent actions and judgements. Although this affective system is poorly understood and not well researched in the psychological literature, Ittelson believes it is central to environmental experience.[96] The aesthetic, as well as the social and systemic components of this ambiance relate to a background of feeling within which aesthetic or other value judgements can be made. Obviously, this is an area requiring a great deal of research and thought, and one much neglected by students of landscape perception at present.

It is evident from this comparison of landscape perception research with Ittelson's psychological approach, that no one of the four paradigms adequately encompasses his notions of how environments are perceived. Yet all make contributions in one area

---

88. Zube and Pitt, op. cit.; Penning-Rowsell, op. cit.

89. D.L. Jacques, "Landscape Appraisal: The Case for a Subjective Theory," *Journal of Environmental Management* 10 (1980): 107-113.

90. N.L. Wilkinson, "Women on the Oregon Trail," *Landscape* 23 (1979): 43-47.

91. See Lowenthal, 1978, 1968, op. cit.

92. Lowenthal, 1979, 1977, op. cit; D. Lowenthal, "Past Time, Present Place: Landscape and Memory," *The Geographical Review* 65 (1975): 1-36.

93. See Relph, op. cit.; Newby, op. cit.; Aiken, op. cit.

94. S. Kaplan, 1979, op. cit.

95. Wohlwill, 1976, op. cit.

96. W.H. Ittelson, Personal Communication, 1982.

token, all four paradigms can be fitted together into a transactional model if some common relationships exist to provide a basis for such an amalgamation. These relationships among paradigms are discussed, briefly, in the following section.

Towards an Integrated Approach to Research

Each of the paradigms makes a unique contribution to the landscape perception literature. There are also, however, several apparent relationships between and among the paradigms that suggest opportunities for the development of a transactional model. The expert and psychophysical paradigms emphasize applications while the cognitive and experiential paradigms emphasize the meanings of landscapes. Interaction outcomes in the expert and psychophysical paradigms tend to be unidimensional and judgemental in nature. Also, the expert and cognitive paradigms share some common definitions of compositional qualities that define valued landscapes, including complexity or variety, mystery, and the importance of edges, for example the junction of field and forest or land and water.

The paradigms also vary in fairly consistent ways with respect to the human and landscape components of the perception model. From expert through psychophysical and cognitive to experiential, the concept of the human varies from that of passive observer to that of active participant. In like manner, the concept of the landscape varies from dimensional in expert and psychophysical to holistic in experiential. Concern with individual landscape resources (topography, vegetation, water) merges into compositional categories or characteristics (mystery, prospect, hazard) and finally into an undifferentiated holistic view of the landscape.

These relationships and consistent variations between and among paradigms suggest potential for development of an integrated framework for future research. Such a framework should accommodate the applications emphasis of the expert and psychophysical paradigms while contributing to our understanding of the meanings of landscape and the nature of human-landscape interactions and interaction outcomes.

Figure 4.3 presents a suggested approach for integrating the study of interactions and relationships between and among humans and landscapes as defined in the four paradigms. This approach draws upon the full array of methods employed in the four paradigms and is intended to search out valid, sensitive and reliable outcomes that have utility for planning and management decision making and that contribute to our understanding of the significance and meaning of landscape to the quality of life.

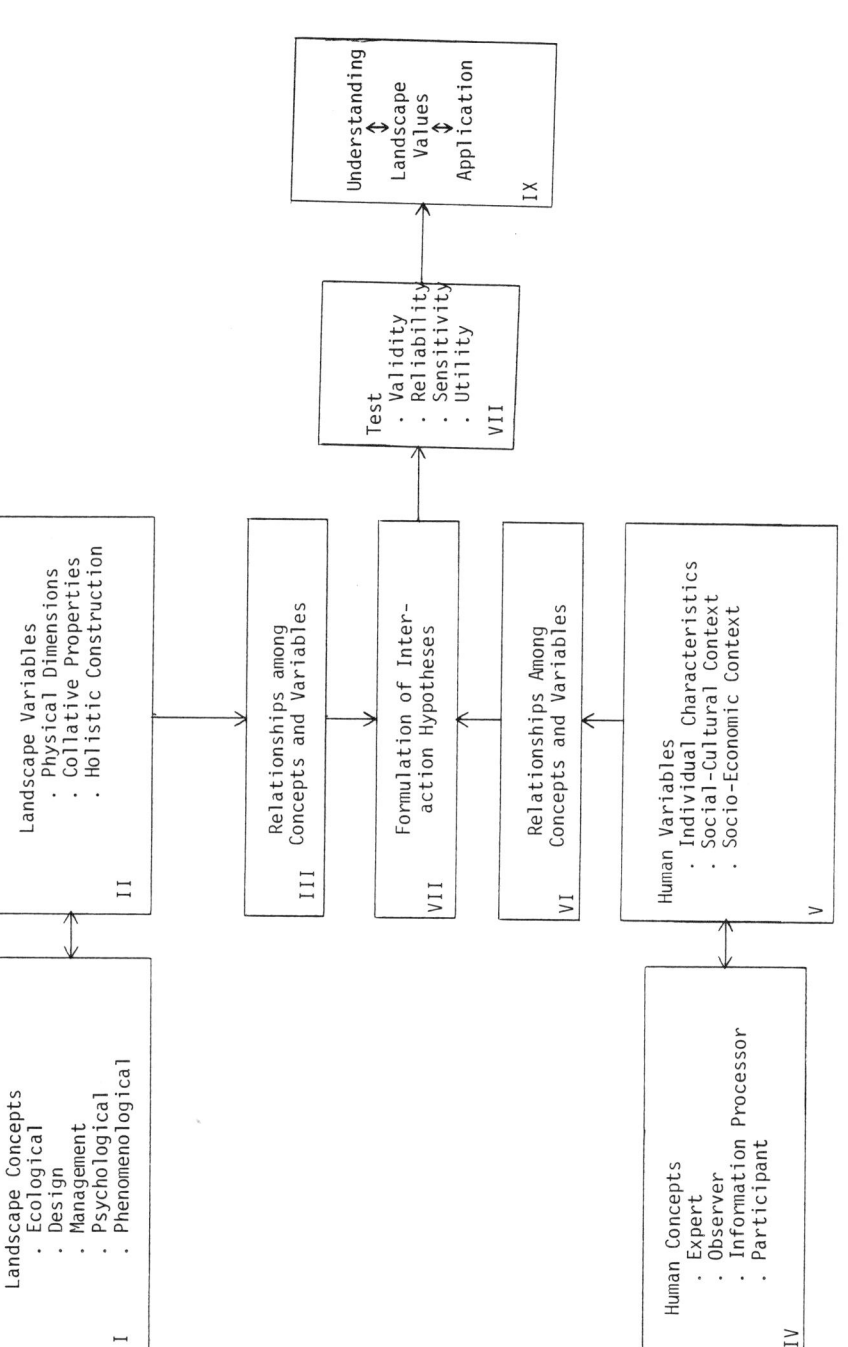

Figure 4.3 Integrated Framework for Landscape Perception Research

This approach is not intended as a single mega-research undertaking but rather as a framework within which individual studies contribute a shared goal of linking understanding and application through theory. Individual research projects can be initiated through questions of either landscape concept and variable relationships (I-II-III) or human concept and variable relationships (IV-V-VI). Whichever avenue of approach is selected, the formulation of interaction hypotheses requires the selection of some concept variable construction from the other. For example, few studies to date have employed the information rich observation or case study methods of the experiential paradigm as an initial means of exploring perceived and experienced values of the landscape. It seems likely that this approach could contribute to the identification of more sensitive human and landscape variables and interaction outcomes and to the formulation of hypotheses to be tested using the more scientifically rigorous methods of the psychophysical and cognitive paradigms (III-IV-V-VI-VII-VIII). Some of the richness and detail are likely to be lost in this process but the results are more likely to be widely accepted and used.

Other paradigm relationships to be studied include:

- between and among concepts of the landscape, for example, are there identifiable and significant relationships between physical dimensions (topography, water, vegetation, etc.) of the psychophysical paradigm and the information content or collative dimensions (mystery, complexity, legibility, etc.) of the cognitive paradigm?

- between and among concepts of the human, for example, what effect does the role of the human--as expert, observer, information processor or participant--have upon the range and salience of perceived values?

- between and among interaction outcomes, for example, what are the relationships among perceived sense of well-being, beauty and habitual behavior in the same and different landscapes?

- between interaction outcomes and social variables, for example, what are the effects of education, group experience, stage in life cycle, etc., upon perceptions and experiences in different landscapes?

Summary and Conclusions

Drawing upon findings from a previous review of journal and related literature[97] we have suggested that there is a compelling

---

97. Zube, Sell, and Taylor, op. cit.

need for an integrative approach to research that can contribute to development of a theoretical framework for landscape perception. Furthermore, this framework must encompass conceptual and applied interests. Indications are that, in the decade ahead, the significance of landscape quality to the quality of life and human well-being will have to be established if it is to remain as a significant and viable policy issue. For quality to remain an important issue, along with economic and other environmental factors in landscape planning and management decisions, it will have to be shown to be important within a hierarchy of social values as it contributes to improving the human condition. In striving for this, we suggest that a transactional view of human-landscape relationships provides a useful integrative model for theory building and that the concepts and methods from heretofore essentially isolated research paradigms be exploited interactively in the development of a transactional theory of landscape perception.

CHAPTER 5
ENVIRONMENTAL PERCEPTION IN GEOGRAPHY
A COMMENTARY

Martyn J. Bowden

The four papers, although all ostensibly dealing with the "perception" of the environment (landscape), inevitably treated that word in four different ways. All of the definitions can be encompassed in what Gold calls behavioral geography (see table 5.1). This is, to him, synonymous with environmental perception and behavior in the broadest sense. In the four papers the authors recognize five facets of and approaches within behavioral geography: (1) spatial cognition; (2) environment-behavior-design; (3) environmental cognition in hazard research; (4) cultural geosophy; and (5) existential phenomenology. In a short space Gold includes all of these but concentrates most on cultural geosophy and spatial cognition, largely because he is dealing with French geography.

Saarinen equates environmental perception with the environment-behavior-design field and with environmental perception in hazards. The perceptual bias in his method of recognition of contributions leads him to de-emphasize, if not exclude, much if not most produced in at least two of the other "facets." Mitchell, who defined his purpose and object precisely and well, restricts himself to hazards research and to the place of environmental perception and behavior research within this field. Sell et al. deal with the landscape tradition in geography as a subset of the ecological-environmental tradition. Sell et al.'s four "paradigms," and the five to which they refer in Daniel and Vining, are shown in the table. I take their psychophysical "paradigm" to be equated largely with spatial cognition; their cognitive paradigm to be equivalent to environment-behavior-design; their expert paradigm is geosophy: aesthetic, cultural, humanistic; and the experiential paradigm is existential phenomenology. The Sell group touches on all but one of the five facets of environmental perception research, strongly emphasizing the two from the Social Sciences in which they work--the cognitive and psychophysical--and using, slightly misusing and possibly abusing the two in which they don't: those from the Humanities and Arts--the expert and experiential.

TABLE 5.1

BEHAVIORAL GEOGRAPHY

Traditions in Geography	Location	Location Env/Ecol.	Ecological/ Environmental	Landscape Region	Place Landscape
5 FACETS of Perception	Spatial Cognition	Environment-Behavior-Design	Environmental Perception in Hazard Research	Geosophy, Aesthetic Historical	Existential Phenomenology
GOLD	② (3)	③			④
SAARINEN		①		① (4)	
MITCHELL			②		
SELL ET AL.	①	②	①	③	(4)
SELL'S "Paradigms"	Psycho-physical	Cognitive ----------- Expert			Experiential
DANIEL AND VINING	Psycho-physical	Psychological		Formal Aesthetic Ecological	Phenomenological

Numbers refer to discussant's assessment of each writer's ranking of the five facets of perception in geography. ②indicates critical facet;②peripheral;(4)highly marginal

Different definitions of environmental perception lead to apparently contradictory conclusions. Saarinen, who sees geographic research in environmental perception as part of "an advancing research frontier at the nexus of environment-behavior-design," sees this as a field and environmental perception as a sub-field of it. In turn, "one of the original subfields of environmental perception in geography [is] natural hazards research." Clearly, to Saarinen, environmental perception is a vast sub-field and natural hazards research is a part of it. He writes of "floods" of articles in the last few years, and "a continued high level of production of research on environmental perception themes." Mitchell, by contrast, sees environmental perception as a "branch," "component," "concept," a small and relatively declining sub-field of the large hazards research field constituting only 15 percent of articles in the last decade. Using a precise definition of environmental perception he recognizes in the last ten years a meagre (30/40) significant articles.

To both Mitchell and Saarinen, environmental perception is nevertheless a sub-field, and it clearly is to Sell et al., who recognize four sub-fields/disciplines in perception (of the landscape) research, each with its own paradigm. Sub-field, sub-discipline, paradigm are to my mind inappropriate terms. I agree with Gold that environmental perception/behavior or behavioral geography--is a dimension omitted from explanation in American geography before the 1960s. Just as the historical approach to geography is called historical geography, so the perceptual, cognitive, behavioral approach to geography is called behavioral geography. To get geographers and others to add this dimension to their arsenal, we--the few--began a geographical movement in the early 1960s. Just as there had been a movement in quantitative geography which became a "revolution," so the movement called elsewhere the cognitive renaissance became the behavioral revolution.[1] Geography as a social and physical science accepted and underwent the quantitative revolution, and American geography as social science and humanistic discipline has thoroughly accepted and is still undergoing the cognitive/behavioral revolution (this is the larger message of Saarinen's paper). As Saarinen correctly states, the techniques of environmental perception "have become thoroughly integrated into geographic research just like the wave of quantitative techniques which preceded them."

---

1. Martyn J. Bowden, "Cognitive Renaissance in American Geography: The Intellectual History of a Movement," *Organon* 14 (1980): 199.

Periods of Vitality in Cognitive/Perceptual Studies in Geography

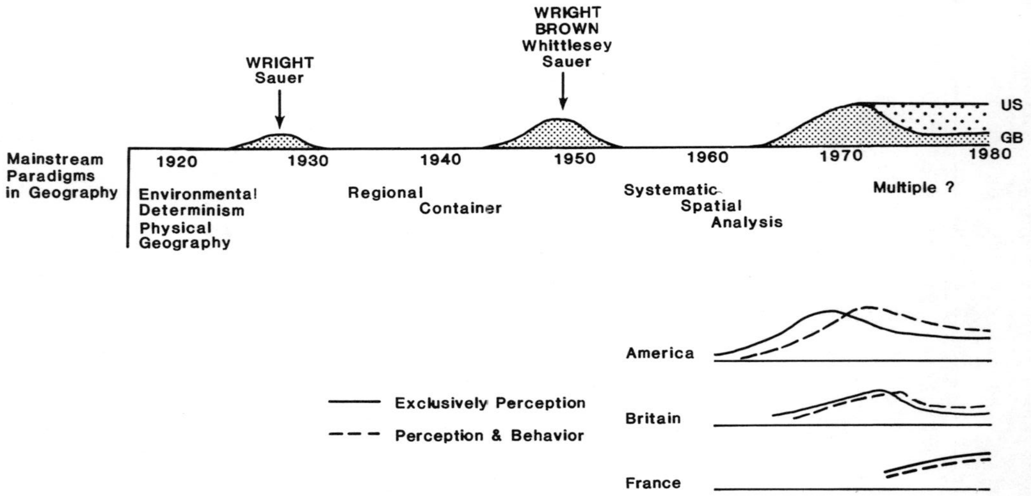

Mainstream Paradigms in Geography	1920	1930	1940	1950	1960	1970	1980
	Environmental Determinism Physical Geography	WRIGHT Sauer →	Regional Container	WRIGHT BROWN Whittlesey Sauer →	Systematic- Spatial Analysis		Multiple ? US GB

America

Britain

France

——— Exclusively Perception
- - - Perception & Behavior

There is a message in this comparison. Just as the quantitative movement (and its methods) were accepted and integrated particularly into the field of urban and economic geography, so the perception movement (and its method) is fusing with and ostensibly disappearing into the urban studies/design (environment-behavior-design) and environmental management (hazards) fields. The meaning of this message is clearly seen in Mitchell's paper. If one takes Mitchell's strict definition of environmental perception there is no reason to expect as Saarinen does "a continued high level of production of research on environmental perception themes." Rather, the perception approach will be present in much research in all of human geography's major fields, but specific specialist articles on environmental perception can be expected to decline in number. For example, in the hazards research field, to which Mitchell devotes most of this paper, environmental perception is now a minor element still dominated by the conviction of the relation between perception and hazard behavior. It is a sub-field in which old established methods and theories are still favored and where investigators in the last eight years recorded "significant," progress on only two of the perception components of Mileti's seven-component model--in hazard awareness studies and in perceptual bias studies. Things are clearly slowing down as they did in quantitative geography.

To claim that environmental perception is a sub-field, as environmental psychology is to psychology, is to my mind divisive and schismatic. It produces in geography, territory and boundary problems that Mitchell rightly deplores. It demands papers such as Sell's dedicated to integrating four "sub-fields" whose practitioners no longer communicate across the broken bridge of C. P. Snow's Two Cultures, but who surely did so in the middle 1960s in the exciting exploratory years (before the territory dividers got to work in the late 1960s).

Gold traces the critical schism in Britain to 1973 when the environmental psychologists created their own subdiscipline and committed themselves to contract work. The rift between the social scientists and humanists in British geography soon followed, as did the ritual denunciations across the bridge, to which Gold as social scientist adds another salvo. In America the government grants-game aided or promoted fractionation, institutionalization, and the strict definition of sub-fields, with the environmental managers (and with NSF) first on the funding train and subject to its demands. The space (and HUD) cadets with the NSF (Social Science) division followed; and humanist/cultural geographers came a late third with NEH, if at all. It is this violent split between the "sub-fields"/paradigms and particularly between the social sciences and the humanities/arts in American geography that Sell et al. essay to repair in their paper (at least in landscape "perception" studies).

They fail because of a social science bias antithetical to their synthetic intent. Theory is the object of the social sciences, and social scientists work under paradigmatic umbrellas, but the quest of the humanities is the idea/logic/and meaning. The humanities/arts--and they dominate two of the four approaches to landscape perception recognized by Sell et al.--are antiparadigmatic/antitheoretical and it is not conceivable to me that the humanists and artists within Sell's "expert" and "experiential" approaches would deny their gods and sell out now to the reductionist methods of the social sciences. The concepts of place and landscape are the outposts of humanism in geography, zealously guarded from the cohorts of *homo economicus optimus* and from the spatial mainstream during its twenty-year domination of American geography. They are not likely to be abandoned easily. Would Sell et al. be willing to abandon their gods, and focus centrally in their work on the idea and meaning of landscape?

The most interesting date that recurs in the papers is 1972/73. In the history of surges of interest in cognition/perception in American geography--that between 1965 and 1970/71 was the third (earlier ones occurred in periods of paradigm uncertainty and change in the late 1920s and the late 40s).[2] In Britain after the euphoria of the exploratory phase, the promise of take-off was not fulfilled, and in 1972/73 the behavioral movement stalled. (Gold in a pessimistic déjà vu sees the same as likely to happen in the French experience.)

Gold suggests that in Britain the beachhead of environmental perception/behavior was initially limited to four centers in a discipline dominated by mainstream spatial analysis. Decline resulted in Britain in 1972, for the same reason that it did in America in 1930 and in 1951/52: isolation in a period of an antithetical paradigm in the mainstream. Gold also adds with insight the lack of new channels of communication and of new avenues for collaboration.

---

2. Bowden, loc. cit., pp. 199-204.

Yet it seems to me that the situation was similar in America in 1972/73. The spatial group was more entrenched than it was in Britain, and "perceptions studies" as Mitchell and Saarinen show were strong in only three/four centers. Why then the continued success in America after 1972/73 in both quantity and quality? What is the difference?

Saarinen points to institutionalization as critical and there is an element of truth in that. Unfortunately, the term suggests to me fossilization, administration, and the dead hand of inertia. I would rather have it that vital centers of research were established, each of them headed and inspired by people remarkably sensitive to each other, and in continual communication with each other, and with kindred disciplines and related government agencies. They were entrepreneurial messiahs, think-tank ideas men, dedicated, flexible collaborators. I refer, of course, to Kates (and his associates at Clark University's CENTED), White and his, at Chicago initially, and later in Colorado, and Burton in Toronto. There is simply no counterpart to this group in behavioral geography anywhere, and I would venture to say in American geography. These are people of stature and maturity who have not relaxed like so many others, after reaching pooh-bah status in their 40s.

There are others too. To Saarinen's hope that the new crop of graduates are our hope for the future, my initial reaction was one of scepticism, until I remembered Gary Moore. As a graduate student in psychology and geography at Clark in the late 1960s and early 1970s, he was a founder and leader of EDRA and was, with other graduate students responsible for a strong outreach to the spatial analysts in America in the early 1970s--a link that bore fruit in a way that did not occur in Britain.

It is, above all, driving, dedicated, messianic, bright, entrepreneurial people, not institutions who effect change and maintain thrust. In their wake there may be floods of material, much of it mediocre and derivative. But within the mass there is quality and innovation as Mitchell so clearly shows in his section on hazard geography and hazard research.

Successful "institutions" need innovative people. It isn't enough to warm over a suggestion made in 1954, and tried again ten years ago. It isn't enough to count articles, write syntheses and review essays, and worry about ritual denunciations of environmental perception. The University of Arizona environmental perception group has the makings of another great and vital center of research. It has the people, it has the challenge, and on the basis of work in the last five years, it has the promise. We must all hope the people succeed, and make that "institution" the fourth geographical success story in environmental perception and behavior in North America.

CHAPTER 6
ENVIRONMENTAL PERCEPTION AND ITS USES
A COMMENTARY

Gilbert F. White

The papers in this volume are strong in their exploration of the philosophical bases for studies of environmental perception, and in reviewing the recent growth in number and domain of those studies. They thereby advance thinking on that topic. They do not come fully to grips with two concerns that motivated the organization and publication of the 1965 symposium at Columbus. These were the efforts to (1) find explanations for the ways in which people cope with environmental risk and (2) develop improved methods for describing perceptions of environmental qualities and options. Believing that those earlier concerns are still relevant, I suggest that the time is right for more systematic appraisal of what has been learned in those respects since 1965 from the uses of such research and from the wide array of methods in geography and associated disciplines.

Between them, Bowden and Buttimer review the several philosophical approaches in ways that are not fully reconciled but that cover most of the views expressed in geographical circles. Except for brief mention in Lowenthal's introduction and in Tuan's paper, this dimension of perception studies received little attention in the Columbus symposium. The focus at that time was on methods that might help illuminate resource management and urban planning in practice.

As I recollect, the immediate inspiration for the Columbus discussions as organized by Kates was the feeling among those working at Chicago on natural resources problems that if they were to find better solutions the description and interpretation of people's views of land use required new modes. It had become increasingly evident since the initiation of flood studies nine years earlier that understanding of how flood plain residents made choices about land and structures involved some kind of insight into how the residents and public officials viewed the resource and its hazards. The social constraints upon individual action had been the target of considerable debate: the effects of property ownership on flood plain speculation, the consequences of government flood protection programs in promoting uneconomic uses, and the inhibiting results of Federal relief on private choice were among the issues in controversy. The actions taken by individuals could not be explained solely from economic analysis assuming optimal choice or as a product of public constraints. Hence, the curiosity as to how people saw the hazards and benefits of flood plain use and its alternatives. Kates had explored this question for flood plain dwellers.[1]

A companion concern was with appropriate techniques of describing the individual's views of environmental phenomena. Differences among what were loosely termed "perceptions" had long been recognized by geographers but there were no widely practiced or tested methods for specifying them. The second aim therefore was to canvass a variety of methods in order to judge their suitability and limitations. Interest centered on approaches developed outside of conventional interview procedures or analysis of literary descriptions.

It was recognized that each of these concerns was far wider than the flood plain. The same questions of process pervaded other segments of natural resources management or urban planning. Methods applicable to flood plains were likely to be applicable to wilderness or soil use or city streets. Saarinen was then using a method borrowed directly from clinical psychology to measure perception of drought hazard.[2] Lucas was seeking an understanding of what "wilderness" meant to different people.[3] Accordingly, no effort was made to limit the subjects covered by the invited papers.

Against this background, it is pertinent to ask how much and in what way the processes and consequences of land use or similar problems are better understood as a result of the studies since 1965, and what lessons have been learned as to the comparative merits and demerits of the available means of studying environmental perception. Only a few of the papers are directly responsive to these concerns. The turn of interest is in itself a significant commentary on the course of studies in the interim: philosophical orientations take precedence over empirical findings. This is symptomatic of broader currents of thought within the geographic

---

1. Robert W. Kates, *Hazard and Choice Perception in Flood Plain Management,* Research Paper no. 78 (Chicago: University of Chicago, Department of Geography, 1962).

2. Thomas F. Saarinen, *Perception of the Drought Hazard on the Great Plains,* Research Paper no. 106 (Chicago: University of Chicago, Department of Geography, 1966).

3. Robert C. Lucas, "Wilderness Perception and Use: The Example of the Boundary Waters Canoe Area", *Natural Resources Journal* 3 (1964): 394-411.

profession. To some extent the change also represents dissatisfaction with the earlier formulations of aims and needs as being fuzzy or insensitive to social structure or unduly positivist, and such continuing critical review is to be welcomed.

With the exception of the paper on natural hazard studies by Mitchell, the review of landscape by Sell, Taylor and Zube, and the pithy comments by Bowden, little is said about the impact of perception studies on the deepening of explanation of geographic processes. Are we any better able because of the work accomplished to analyze rural or urban land use than we were in 1965? The evidence is slim for fields other than natural hazards and only a few bits are noted in this volume.

So far as natural hazards are concerned it appears that the introduction of a perceptual dimension has led to somewhat more satisfactory statements in a few sectors, such as how people respond to warnings and how they decide to buy or not buy insurance and the circumstances in which they take steps to mitigate the deleterious effects of extreme events. The simplified explanations prevailing in the early 1960s are no longer accepted. At the same time the notion of perceived risk has found wide application in the burgeoning field of technological risk analysis, and it is commonplace to hear experts in a wide range of fields distinguish between "real" and "perceived" risk. Some use of geographic studies of perception has been made in the field of domestic water management. Forest and park managers involved in recreational use, particularly of wilderness areas, employ the concept routinely. It is increasingly mentioned in the whole arena of environmental design and behavior. Other uses might be cited. Just how influential a role geographers have played in these developments might be ascertained.

This is not the place to venture a comprehensive review of the uses or failures of the perception concept in studies or planning having to do with the environment. It is the place to propose that such a survey would be timely and should be undertaken. Without it, any judgment as to the significance of the work accomplished is bound to be incomplete. And it could be linked with a more systematic appraisal of the techniques employed.

This volume was intentionally restricted by the editors to the work of geographers whereas the earlier volume reached out consciously to other disciplines for experience with observational methods. A reading of certain of the papers and particularly of the reviews in Progress in Geography by Saarinen and Sell[4] reveals a wide range of experimentation and a willingness to look to neighboring fields for methods. The array is highly eclectic, and while earlier enthusiasm for trying a variety of techniques may have subsided, there is little evidence of pedestrian acceptance of a few methods. However, there may not have been sufficiently critical examination of the experience with those diverse modes of investigation that have been tried. Critical reviews have been undertaken in a few related fields, for example, Heberlein's appraisal of social psychological studies of environmental attitudes.[5]

It is noteworthy that apparently there has been no systematic comparison by geographers of the results from applying and replicating a variety of methods to the same set of phenomena. Would the findings be greatly different, and if so how, were reliance placed on individual interviews, non-directive interviews, content analysis, or any of the score or more types of tools that might be applied. And how much confidence can there be in studies which have not been replicated? This suggests the second step that might now be taken toward a fuller appraisal of what has been learned about method since the environmental perception concept began to gain popularity. Until this is achieved the answer to the question of relative effectiveness will remain speculative.

---

4. Thomas F. Saarinen and James L. Sell, "Environmental Perception," *Progress in Human Geography* 5 (1980): 525-547.

5. Thomas A. Heberlein "Environmental Attitudes," unpublished paper, Department of Rural Sociology, University of Wisconsin, 1980.

PART II
REVIEWS OF RESEARCH
URBAN AND MICROSCALE THEMES

CHAPTER 7
THE GEOGRAPHY OF CHILDREN AND CHILDREN'S GEOGRAPHIES

Roger Hart

Introduction

Whether geographical behavior begins with a baby's first exploration of its own body, its spatial experiments in calling for its parents' attention, or when it first crawls away from the "nest," it must surely be agreed that human geography begins in childhood. Why then has so little been written by our profession on the geography of children? Clearly research questions in relation to them have been considered trivial. Children have until recent years been ignored as a suitable population for basic research, leaving them at that small, and rather isolated, branch of geography called geographic education.[1] This is in error, for our greatest period of geographical exploration and learning is that found in each of us, in our childhood. This paper is designed to illustrate some of the developments in the study of children since the beginning of "behavioral geography" approximately fifteen years ago.

Beyond the continuing research on geographic education which is almost entirely directed to the formal setting of school classrooms, it is now possible to identify two areas of basic research with children which are unequivocally geographic even if much of the work is not being done by geographers. First, there is a growing body of literature on the environmental behavior of children, particularly concerned with spatial activity and use of the landscape. Much of this research has been conducted by planners, but it is nevertheless geographic. Second, there is an expanding literature on the development of children's knowledge of the geographic environment, most of it being focused on children's understanding of the spatial location of phenomena in the landscape. Although geographers have played a key role in igniting this

---

1. B. S. Bartz, "The Nature of Research in Education," *Journal of Geography* (April, 1972); N. J. Graves, *Geography in Education* (London: Heinemann, 1975).

research direction, the fire is being fed mainly within developmental psychology. I will treat each of these two fields in turn, reviewing what kinds of geographic questions have been addressed, in what ways, and some of the exciting challenges for those geographers who would join this research effort.

The Geography of Children

After reviewing psychological theory and research and carrying out a little research myself on children's understanding of spatial relationships in the geographic environment, I realized that for our theories to advance, more needed to be known about children's spatial behavior.[2] I was shocked to discover that not only had geographers ignored this question but that there were no studies of children's spatial behavior in their everyday outdoor environment by any discipline.[3]

Ironically, I discovered at the same time, a monograph on baboon ecology.[4] So much more advanced was the understanding of baboons' spatial behavior than that of human children that these investigators were hypothesizing the relationship of their spatial behavior to their spatial knowledge even though the baboons could not speak or otherwise represent their knowledge! The failure of psychology to have investigated children's geographic behavior was primarily due, I believe, to a methodological determinism. Almost all of the naturalistic studies of children were observations of behavior in the classrooms of the earliest grade levels of school.[5] It is more difficult to observe older children, particularly when they are ranging over large areas of outdoor space, than in the enclosed setting of a classroom. Nevertheless, by reviewing the many separate domains of psychology, nuggets of theory relevant to a geography of children can be found. Some of the more valuable ones are reviewed below.[6]

---

[2]. R. Hart and G. T. Moore, *The Development of Spatial Cognition of Large-Scale Environments*: *A Review,* Monograph of the Center for Human Environments, City University of New York Graduate School, 1976 (originally published Clark University, 1971); R. Hart, *Children's Experience of Place* (New York: Irvington, 1978), introductory chapters.

[3]. I was aware of a pre-war German study: M. Muchow and H. Muchow, *Der Lebensraum des Grossstadtkindes* (The Life Space of the Child in the Large City) (Hamburg: Verlag, 1935). But this is only now being translated (see *Childhood City Newsletter,* 1981, for progress report).

[4]. J. Altman and S. Altman, *Baboon Ecology* (Basel, Switzerland: S. Karger, 1968).

[5]. H. F. Wright, "Observational Child Study," in *Handbook of Research Methods in Child Development,* ed. P.H. Mussen (New York: Wiley, 1960).

[6]. For a more comprehensive account see the appendices in Hart, op. cit., 1978.

Through the observation of infants and interviews with children's parents, Gesell and colleagues[7] gave us a straightforward account of the normative phases of children's behavioral development, one of which was locomotor development, including the use of different means of transport. While of documentary value, this research tells us little of theoretical interest to the geography of children; it is a simple average age-related charting of the occurrence of behaviors.

I found some psychoanalytic speculation of interest,[8] but the empirical research of John Bowlby on the attachment behaviors of mother and child was the most relevant.[9] Through naturalistic observations of mother-child interaction, Bowlby demonstrated how the child's gradual exploration away from the mother is a negotiative developmental process for both parent and child. Applying this model to the environment with home as the secure base, it is clear that one needs to study not only child development but child-with-parent development for a complete understanding of a child's gradual engagement with the geographical world. For example, it has been found in experimental situations that children will explore more readily if their mother is present, a difference which becomes particularly marked when a strange person is present or when mother and child are in an unfamiliar place.[10]

After arguing for the importance of children having a permanent home-base, Donaldson and Aldrich[11] in a review article on "Children and the Urban Environment" hypothesize that there is a developmental sequence in which the territorial range at one level of experience becomes the home base for the next stage of exploration: mother as home base for exploring the room is replaced by the room as home base for exploring the house, etc. This is

---

7. A. Gesell, ed., *The First Five Years of Life: A Guide to the Study of a Pre-School Child* (New York: Harper, 1940); A. Gesell, F. Ilg, and L. Ames, *The Child From Five to Ten* (New York: Harper and Row, 1946).

8. See, for example, E. Schactel, *Metamorphosis: On the Development of Effect Perception and Memory* (New York: Basic Books, 1959); H. Searles, *The Non-Human Environment in Normal Development and Schizophrenia* (New York: International Universities Press, 1959); E. H. Erikson, "Toys and Reasons," in *Childhood and Society* (New York: Norton, 1963); B. Bettelheim, *The Children of the Dream* (New York: Avon Books, 1969); reviews in appendices of Hart, op. cit.

9. J. Bowlby, *Attachment and Loss, Vol I: Attachment* (New York: Basic Books, 1969).

10. For example, N. Ainsworth, D. Salter, and A. B. Wittig, "Attachment and Exploratory Behavior of One Year Olds in a Strange Situation," in *Determinants of Infant Behavior*, ed. B. V. Goss, vol. 4 (London: Methuen, 1969); H. L. Rheingold, "The Effect of a Strange Environment on the Behavior of Infants," in Goss, op. cit.

11. O. F. Donaldson and R. A. Aldrich, "The Child in the Urban Environment: A Review of Literature and Research," in *42nd Yearbook of the National Council for the Social Studies*, ed. R. Wisnewski (Washington, D.C.: National Council for the Social Studies, 1970).

similar to the developmental sequence of space needs theorized by the planner Doxiadis.[12] Margaret Mead[13] in an article on "Neighborhoods and Human Needs" argued, however, that a permanent home base is not necessary to establish the "basic need" for continuity. She found through her extensive travels with her young daughter that the simple placement of the potty in a new room was enough to establish "home". Similarly the family car can be enough for many children when moving to a strange place. It is strange that this question of the role of the physical environment in helping establish a sense of permanence for children is not discussed more in the literature of clinical psychology.[14] A related area of theory, on the importance of "transitional objects" has been explored, however: Christopher Robin's teddy bear, Winnie the Pooh, is the most famous of these special objects, but it is well known by parents in general that a few months after birth, infants are likely to become attached to some particular object or furry toy.[15] Winnicott,[16] a psychoanalyst, has incorporated this phenomenon into theory on the child's struggle to bridge the gap from the egocentricity of early infancy to the recognition of an external world of adults with independent actions. A transitional object is, he believes, the first unchallenged area of experience which is neither the self or the mother. The most obvious geographical relevance of this theory is that a transitional object, being effective in reducing anxiety related to separation from the mother, can be carried around and serve as an aid in the child's exploration away from the safe nest of the crib.

One group of researchers, the ecological psychologists, have for many years observed children in their everyday geographical-scale environment.[17] Unfortunately, in spite of the promise of the term "ecological," these psychologists have not paid much attention to the spatial and physical properties of children's

---

12. C. A. Doxiadis, *Anthropopolis: City for Human Development* (New York: Norton, 1975).

13. M. Mead, "Neighborhoods and Human Needs," *Ekistics* 123 (February 1966), pp. 124-126.

14. Reviewed in Searles, op. cit.

15. O. Stevenson, "The First Treasured Possession: A Study of the Part Played by Specially Loved Objects and Toys in the Lives of Certain Children," *Psychoanalytic Study of the Child,* vol. 9 (New York: International Universities Press, 1954), pp. 198-217; J. Newson and E. Newson, *Four Years Old in an Urban Community* (Chicago: Aldine, 1968); J. Newson and E. Newson, *Seven Years Old in the Home Environment* (Harmondsworth, Middlesex, England: Penguin, 1978).

16. D. W. Winnicott, *Playing and Reality* (London: Tavistock, 1971).

17. R. G. Barker and H. F. Wright, *One Boy's Day* (New York: Harper and Row, 1951); R. G. Barker and H. F. Wright, *Midwest and Its Children: The Psychological Ecology of an American Town* (Evanston, IL: Row Peterson, 1955); R. G. Barker, *Ecological Psychology: Concepts and Methods for Studying the Environment of Human Behavior* (Stanford: Stanford University Press, 1968).

worlds. Like most other psychologists, the emphasis has been upon observing children's social transactions. There are exceptions, however, for sometimes the influence of the physical environment cannot be ignored.[18] The importance of the physical environment becomes more clear when one allows children to represent their worlds through some medium instead of simply relying upon observation of them. A study of children's behavior in, and representation of, a small town in comparison to that of child residents of a large town led to some very provocative findings by the ecological psychologists.[19]

There are many studies of child-rearing which use spatial restrictedness as one index of socialization practice.[20] All of them rely upon interviews of parents as their data source and the results are usually superficial. An exception is the remarkable series of books by psychologists Newson and Newson reporting different phases of their longitudinal research of child-rearing in Nottingham, England.[21] By using open interviews and becoming very familiar with the child-rearing philosophies of individual parents they were able to balance and enrich the more traditional reporting of aggregate data with detailed descriptions from the parents themselves. Although children's spatial behavior and relation to the environment were not major foci of the research, in each of the books they emerge frequently as important factors in child-rearing. Particularly interesting are the numerous discussions of how these factors are differently seen by parents of the five different social classes investigated. It becomes clear from this research that not only objects, such as toys, but characteristics of the larger physical environment, including its spatial arrangement, are important in influencing the socialization of children.

There is one valuable piece of research carried out by a geographer specifically on children's spatial behavior.[22] An

---

18. P. Gump, "Behavior of the Same Child in Different Milieus," in *The Stream of Behavior,* ed. R. Barker (New York: Appleton Century Crofts, 1963), pp. 109-222.

19. H. F. Wright, "Children in Small Town and Large Town U.S.A.: A Summary of Studies" (Department of Psychology, University of Kansas, mimeo, 1979).

20. For example, J. W. M. Whiting and I. I. Child, eds., *Six Cultures: Studies in Child Rearing* (New York: Wiley, 1961); D. Landy, *Tropical Childhood* (New York: Harper and Row, 1965).

21. J. Newson and E. Newson, *Patterns of Infant Care* (Harmondsworth, Middlesex, England: Penguin, 1963); Newson and Newson, op. cit., 1968, 1976; W. M. Wiest, "Children's Situations in Communities Differing in Sizes as Revealed by a Projective Test of the Environment" (Master's thesis, University of Kansas, 1957).

22. M. Tindal, *The Home Range of Black Elementary School Children: An Exploratory Study in the Measurement and Comparison of Home Range,* Place Perception Report no. 8 (Worcester, Mass.: Graduate School of Geography, Clark University, 1971); J. Anderson and M. Tindal, "The Concept of Home Range: New

interview with second and fourth grade children in an urban ghetto and a suburban area of Baltimore used an aerial photograph to help obtain information on the child's "home range." In accord with her hypotheses, the ranges of the suburban children were larger and the ranges of the fourth grade girls in the suburban environment were less than those of the second grade boys. Speculations were made to interpret these findings but more wholistic or ecological studies involving an in-depth understanding of the environment itself and other actors in it, particularly the parents and other child-caretakers, is required in order to provide a comprehensive account of the dynamic set of forces involved.

In an attempt to inform and build theory across the traditionally divided domains of child-environment research--spatial activity, spatial cognition, environmental perception and preference, and use of the environment--I conducted a descriptive developmental study of children's place experience in a New England town.[23] By taking an integrative look at both the geography of all of the towns' children and their individual geographies, some contributions to existing theory were made possible. In the realm of spatial exploration it was surprising to find that the most frightening of places in the town were also among the most attractive to them; an observation not so surprising when one recalls one's own childhood or some of the more successful children's books such as *Huckleberry Finn*. A more important discovery was the necessity of thinking of children's spatial ranges as the product of negotiation between parent and child. Planners and designers have frequently failed to recognize the interactive nature of child-parent (or caretaker) relations in their housing guidelines, particularly in the many critiques of high-rise housing (e.g., Department of Environment, 1973).[24]

These professionals write of the need for mothers to be able to watch over their children, as though children themselves feel no need to maintain contact with a caring adult; parents, in fact, commonly set spatial range rules around the needs and demonstrated readiness of their children, rather than requiring the children to meet their rules. Seen this way, the solution is not to simply design housing areas to guarantee the abilities of parents to watch over their children.

Data for The Study of Territorial Behavior," in *Environmental Design: Research and Practice,* Proceedings of the Environmental Design Research Association, Third Conference, ed. W. J. Mitchell (Los Angeles: University of California, 1972).

23. Hart, op. cit.

24. Department of the Environment, *Children at Play* (London: Her Majesty's Stationery Office, 1973).

Confirming the findings of Tindal, described above, I found boys' spatial ranges to be significantly larger than those of girls. It is remarkable that such dramatic sex differences in experience had not been noted by child psychologists--a reflection of the fear of research beyond the laboratory, and the preoccupation with age as the critical variable. Opportunities to manipulate the physical environment were also much less for girls in the town. I discovered that toys, tools, play equipment and landscape qualities were seen by parents as important "tools" in their children's socialization. This became more obvious to me as I began to notice social class differences in the design and layout of space around the homes and in the provision of toys. Related to different ideologies and philosophies of child-rearing, the home landscape looked completely different in the various neighborhoods of the town. The manual-working families allowed their children to use the landscape around their home freely in their play. They commonly encouraged resourcefulness in the use of materials by buying them tools and other "working" toys such as pulling wagons or fishing rods. In contrast, the landscape of the middle-class professionals' homes was manicured and controlled to such a complete degree that their use of the environment was limited to specific sites and specialized toys and equipment. In an unsystematic manner I was observing the use of the physical environment as a tool in socialization; this is a valuable unexplored area for pursuit by any geographer interested in the role of the environment in social and cultural reproduction.

Most other research has been done by urban planners, landscape planners and architects, sometimes with the assistance of sociologists.[25] Unfortunately, the research by planners is usually of the observational survey type and hence suffers from superficiality--observers simply record the location, sex and approximate age of children and their activity according to some gross category of play. Little or nothing can be said from this research about the children's spatial behavior because their identity and hence home location is not known. Also, etic categories of landscape and behavior are used rather than the children's own categories.

---

25. Ibid., in which there is a summary of the most comprehensive of these; Hart, op. cit.; R. Moore and D. Young, "Childhood Outdoors: Toward a Social Ecology of the Landscape," in *Children and the Environment,* ed. I. Altman and J. F. Wohlwill (New York: Plenum, 1978).

Other research, also motivated by a desire to influence planning policy and environmental design for children, has used interviews instead of, or in addition to, observation.[26] This research integrates this section of the paper with the second section on "Children's Geographies," but its impact has been greater on environmental planners and designers than on psychologists and educators and so it is primarily discussed under this first section. The research includes children's knowledge of landscapes, their accessibility to and use of places, and their evaluation of them, primarily in terms of their suitability for play. Unfortunately most of this research is of the case study variety, and hence is of limited utility in guiding planning and design. Two recent exceptions are by planners: the comparative work of Robin Moore[27] on children's perception and use of three very different English landscapes, and Kevin Lynch[28] in a UNESCO-supported study of children's landscape perception and use in four cities in Australia, Argentina, Poland and Mexico. The more recent research by Lynch also included broader questions of children's cognition of the social and economic state of their neighborhood and their assessment of their own future in relation to these settings.[29] While very suggestive, none of these studies makes specific links to the objective economic conditions of the children's families. Also, like the rest of our research, it is not clear how they can help guide the way to change in the quality of children's lives. It should be noted that all of this research, including the UNESCO study, suffered from extremely limited budgets; there still has not been, to my knowledge, a funded project of any large scale on the geography of children.

Van Vliet recently completed a study of suburban and urban children's neighborhood perceptions which is a beginning for the kind of comparative research we need.[30] One hundred and sixty-eight children from urban and suburban Toronto responded to forced choice

---

26. M. Southworth, "An Urban Service for Children Based on Analysis of Cambridgeport Boys' Conception and Use of the City" (unpublished Ph.D. dissertation, Massachusetts Institute of Technology, 1970); E. Bussard, "Children's Spatial Behavior in and Around a Moderate Density Housing Development: An Exploratory Study of Patterns and Influences" (unpublished Master of Science thesis, Cornell University, 1974); M. J. Chombart de Lauwe, *Enfant et Jeu* (Paris: Editions du C.N.R.S., 1976); K. Lynch, *Growing Up in Cities: Studies of the Spatial Environment of Adolescence in Cracow, Melbourne, Mexico City, Salta, Toluca and Warsaw* (Cambridge, Mass.: MIT Press, 1977); R. Moore, *Childhood's Domain: Place and Play in Child Development* (London: Croom-Helm, in press).

27. Moore, op. cit.

28. Lynch, op. cit.

29. Ibid.

30. W. van Vliet, "Neighborhood Evaluations by City and Suburban Children," *American Planning Association Journal* 47 (1981): 458-466.

polar adjectives describing their neighborhood and answered open-ended questions concerning their neighborhood and the larger urban environment. The study confirmed some of the expectations generated by the largely less comparative literature of the past. The suburban children, for example, described their neighborhood more often as boring, and safe, and city children more commonly referred to unfriendly people, noise and dirt as disliked qualities of their neighborhood. There were, however, some more intriguing findings. More city children had places in their neighborhood where they were afraid to go than did suburban children, but more of the suburban children were afraid of the larger urban environment beyond their residential neighborhood. The most likely interpretation is that the city children had greater knowledge and competence in facing this wider environment. Such speculation raises exciting possibilities for research. Even more useful findings were achieved when the investigator went beyond the simple comparison of urban and suburban to discover what specific variables in the residential area might be important. The children were re-categorized on the basis of such factors as the density of peers living around the home and the variety of land-use types available in close proximity, and analyses of variance on the children's responses was computed. The findings are interesting: children from neighborhoods with lower densities had fewer friends and complained of a shortage of friends as a disliked feature of their neighborhood; children living in neighborhoods with a greater number of recreation and assembly places and institutional facilities such as fire stations and community centers knew more adults. We need more such thoughtfully designed research to help us discover what specific environmental features are important to children and teenagers. It would be ideal if such quantitative, comparative analyses could be supplemented with more qualitative participant-observer research with a smaller number of the same children. Perhaps then we might move beyond the polar comparisons of these two types of research approach as "hard but irrelevant" and "soft but sensitive" to a self-critical kind of research which fears the possible superficiality of aggregate survey research and avoids the naive generalizing of descriptive case studies, by combining both.

Geographers have traditionally been concerned with large spaces. When I first designed the research in a New England town, I began at the doorsteps of the children's homes. This was probably not a serious error for the New England town I chose to investigate, but for my current research in New York City it would be a gross mistake.[31] It is crucial to understand the relationship of indoor to outdoor use of space by children; the reduction of freedom to use the out of doors in urban areas is probably one of the most important changes in children's lives this century. To understand this change, the reasons for it, and its impact on children, we need to conceptualize their total space use, making no arbitrary scale or territory distinctions: geography merges with architecture.

One underused source of data is literature, including childhood autobiographies.[32] The major problem in turning to literature as a source of insights about children's geographical worlds is that the writers form an extremely biased sample; few of those who have published as adults grew up in very poor families. One alternative is to turn to children's own writings; there are a number of general anthologies.[33] It is regrettable that all of the studies of children's environmental perceptions have shied away from incorporating such self-expressions into their research, no doubt because they are not thought to be part of the battery of accepted methodologies in social science. One unique study, however, asked adults about their childhood memories of their city environment.[34] The research is informal and modest. Quite rightly it begins with a warning not to see memory as simply an accurate but partial photograph of the past but one that has been colored-in and modified over the years. Nevertheless, places were recalled with vivid detail and there were strong similarities, giving the authors confidence that some valid generalizations could be made. When contrasted with the reports of the observational land-use studies described above, it is clear that they better capture the qualities of places which are important to children and also reveal places commonly hidden from an observing social scientist. The study obtained particularly vivid records of places free from adult authority, not specially planned for children and suitable for modification, physically and imaginatively.

William Bunge argued forcibly in the late 1960s and early 1970s for the need to study the environmental forces impacting the lives of what he considered our largest minority--children. He

---

31. Hart, op. cit.; R. Hart and L. Chawla, *The Development of Children's Environmental Concerns* (New York: Center for Human Environments, City University of New York Graduate School, 1981).

32. See "Environmental Autobiographies," *Childhood City Newsletter*, no. 14 special issue, 1978; L. Chawla, *Past Place/Present Time* (New York: Center for Human Environments, City University of New York, Graduate School, 1982).

33. S. Cornish and L. W. Dixon, eds., *Chicory: Young Voices from the Black Ghetto* (New York: Association Press, 1969).

34. A. K. Lukashok and K. Lynch, "Some Childhood Memories of the City," *Journal of the American Institute of Planners* 22 (1956): 145-152.

demonstrated what a research program in the geography of children might look like and what its value might be through a series of small studies in Detroit.[35] Current research lacks the powerful political intentions of Bunge's work. He contrasted not only children's "perceptions" but also the objective conditions of their environments. For example, one simple study was a mapping of negative landscape elements (broken glass, garbage, etc.) in a few blocks of a Black ghetto in the city of Detroit, contrasted with the mapping of the same elements in a similar area of a Detroit suburb. Similarly, other maps show the relationship between the number of traffic accidents involving children and the racial distribution of the population. The investigators also attempted to balance this outside view with "inside" descriptions by teenagers of the environment of the ghetto. The work expanded to Vancouver and later to Toronto, but unfortunately did not come to have the impact on the public which it was meant to, nor did it have much impact on other research. The orientation was original and had the potential of influencing policy-makers in both public participation and information.[36] It probably failed for a number of reasons. While it was overtly more relevant than most other research has been, it was methodologically sloppy and too subject to rhetoric.

A major problem with current research is that it is not clear, even when we find out how children discriminate and use the landscape, what this should mean for planning and design practice. What do we use as our guide for what children should have available for them? The children's statements? The parents' ideas? Or some summary extracted from the child psychology literature? This question is usually not discussed. My own belief is close to that of Bunge. Landscape planning for children should be carried out at the community level where it can be culturally and environmentally sensitive to local demands. The role of research should be to help reveal how the environment is currently used and perceived by children and parents, and to offer some commentary from the collective wisdom of behavioral science but to allow parents, with their own differing values for child-rearing, to play the major role in the decisions, ideally with their children participating.[37]

---

35. Y. Colvard, ed., *Field Notes*: *The Geography of the Children of Detroit,* Discussion Paper no. 3 (Detroit: Detroit Geographical Expedition, 1971).

36. A valuable spin-off from this work is a book for schoolteachers by Tom Scanlan with some of the ideas gleaned from his work with the Toronto Geographical Expedition; T. Scanlan, *Neighborhood Geography*: *A Research Manual* (Toronto: Is Five Foundation, 1978), covers exercises which enable children to look more closely at their own environments and is very close to the tradition of B.E.E., to be described below.

37. See *Childhood City Newsletter,* published quarterly by the Center for

Bunge's approach should be applauded for avoiding the naive idealistic stance of collecting data on children's environmental perception and behavior in the hope that it will be picked up by some concerned planner. His more effective route of achieving local action and reaction by developing research tools to be used in the hands of local residents themselves is being more modestly pursued by geographers and others in the U.K., to be described below.

Children's Geographies

Jean Piaget has unwittingly had the greatest influence on research in children's geographic thinking in North America. It is surprising for many to discover that in spite of his powerful influence in the field of cognitive developmental psychology[38] and education, these disciplines were not his major concern. He was a "genetic epistemologist," an empirical philosopher, concerned primarily with the structure of knowledge, which he believed could best be discovered by investigating its development in young children. As a constructivist philosopher, he believed that what we take to be real is a construction of thought; a product of the interaction between a child and its environment, between maturation and socialization. In order to study epistemology one must therefore proceed developmentally, for only in process is the construction of thought revealed. Thus we have over fifty years worth of clinical experiments with children by Piaget and his colleagues in Geneva on children's thought and language, and their concepts of time, space and causality.

Surprisingly, geographers interested in geographic learning and education have largely ignored Piaget's early books which are particularly geographic: *The Child's Conception of the World*,[39] *The Child's Conception of Physical Causality*,[40] and *The Construction of Reality in the Child*,[41] has much to say that is of relevance for geographic education, but even more important is their value as a prospectus for further research. Their descriptions of the child's conception of the earth, sun, and the moon, the weather, rivers,

---

Human Environments, City University of New York, Graduate School: no. 22 (1981), "Participation: Is It Really Happening?;" no. 23 (1981), "Participation 2: A Survey of Projects, Programs and Organizations;" no. 28 (1982), "Participation 3, Participation: Techniques."

38. Throughout this paper the terms "cognition" or "cognitive" may be thought of as equivalent to thinking or knowing.

39. J. Piaget, *The Child's Conception of the World* (Totawa, NJ: Littlefield, Adams & Co., 1967, original French ed. 1926).

40. J. Piaget, *The Child's Conception of Physical Causality* (Totawa, NJ: Littlefield, Adams, Patterson, 1960, original French ed. 1927).

41. J. Piaget, *The Construction of Reality in the Child* (New York: Basic Books, 1954, original French ed. 1937).

lakes, seas, trees, and mountains stands as an embarrassment to the research in geographic learning that has followed their publication in English over forty years ago. Piaget's own research subsequently moved away from the exploration of children's interests and knowledge of the workings of the everyday physical environment into experimental designs with minimal sets of physical elements. This was done in order to tighten his understanding of children's logical thinking and thereby provide a general developmental account of the structure of children's thought, regardless of the content.

In the 1940s Piaget wrote two additional books of great relevance to developmental geography: *The Child's Conception of Space*[42] and *The Child's Conception of Geometry*.[43] These books have been recognized and used a great deal by geographers and educators, and recently by developmental psychologists concerned with children's understanding of maps and their mental constructions of spatial relations or "cognitive maps."[44]

Through his emphasis on the structure of children's thought, Piaget left behind any concern with the content of their thought, that is, the particular phenomenon being discussed with the child and his or her interest in it, or lack thereof. Susan Isaacs was one of the early critics of this contentless research in terms of its relevance for educators.[45] She felt that it was necessary to investigate children's affective development simultaneously with their ability to logically construct the world, that is, we need also to be concerned with children's interests in the world. This criticism of Piaget's educational relevance remains true today and can equally be applied to the profession of education as a whole. It is for this reason that we have hundreds of individual experiments on the development of the structure of children's spatial thinking[46] and hopelessly few integrated studies of how children explore and understand the large scale spatial environment, what features they selectively engage with and what questions they

---

42. J. Piaget and B. Inhelder, *The Child's Conception of Space* (New York: W.W. Norton, 1967, original French ed. 1948).

43. J. Piaget, B. Inhelder, and A. Szeminska, *The Child's Conception of Geometry* (New York, Basic Books, 1970, original French ed. 1948).

44. Reviews by Hart and Moore, op. cit.; A. Siegel and S. H. White, "The Development of Spatial Representations," in *Advances in Child Development and Behavior*, vol. 10, ed. H. W. Reese (New York: Academic Press, 1975).

45. S. Isaacs, *Intellectual Growth in Young Children* (New York: Schoken Books, 1966, originally Routledge, 1930). See Appendix on "Children's Biological Interests", in S. Isaacs, *Social Development in Young Children* (New York: Harcourt Brace, 1933).

46. Hart and Moore, op. cit.; Siegel and White, op. cit.; J. Elliot and N. J. Salkind, eds., *Children's Spatial Development* (Springfield, IL: Charles Thomas, 1975); L. Liben, A. Patterson, and N. Newcombe, eds., *Spatial Representation and Behavior Across the Life Span* (New York: Academic Press, 1981).

spontaneously ask.[47] It would therefore be a mistake to rely solely on psychology as our primary indication of what a relevant developmental geography should look like. Let us rather turn to the close observations of children in their natural settings as the guide to our design and interpretation of experimental research and to the elaboration of theory.

In 1934 a remarkable little book was published by Lucy Sprague Mitchell called, *Young Geographers: How They Explore the World and How They Map the World*.[48] Lucy Sprague Mitchell has been an important figure in the development of progressive ideas in education. She believed geography to be a very important part of elementary school curricula. Her book contains a table entitled, "Notes Towards a Chart of the Development of Geographic Thinking and Tools." Although she describes this table as "sketchy, superficial and a bit pompous," it represents a clear statement of the direction we as geographers should have been following in studying children's geography. The table charts the progressive stages of children's geographic interests, orientation, and understanding of geographic relations, and the modes used for representing the environment: "lenses and muscles, toys and blocks, picture-drawing, written language and maps." In the half century since this informal charting of observations and ideas of an enthusiastic and insightful educator, we have done little in any systematic or coordinated manner to improve our understanding of the development of children's exploration and knowledge of the geographic environment. There are, however, observations and conceptualizations about developmental geography scattered in a wide variety of places. There is an existing body of theory and method in developmental psychology to draw from and the promise of making important contributions to educational practice.

The need for a research program is particularly great in the United States where geography as an elementary school subject is almost unknown and where map skills are dealt with so poorly that most children do not have a single atlas available in their classroom. The National Council for Geographic Education (unfortunately isolated from the Association of American Geographers) and its periodical, the *Journal of Geography,* provides a forum for educators. The emphasis, however, is entirely upon education in formal, or institutional settings, rather than upon basic research in children's geographic learning.

---

47. Hart, op. cit; R. Moore, op. cit.

48. L. Sprague-Mitchell, *Young Geographers: How They Explore the World and How They Map the World* (New York: Bank Street College of Education, 1934).

Elementary geographic education is nothing if not an interactive process between the curriculum and the geographic interests and understanding of the child. Hence, research in elementary school geographic education should reflect both sides of this interactive process. This is not the case. The bulk of attention continues to be placed on the side of curriculum development based on the structure and content of geography with little regard for the spontaneous development of children's geographic experiences, interests, and thoughts. Educational research and practice still reflect little awareness that children learn outside as well as inside the schools.[49] We need research programs which address children's spontaneous geographic questions. What follows is a review of those few sub-areas of geography which have been considered by the small number of investigators who listen to children.

Physical Geography

Given the crude state of our knowledge of children's understanding of physical geography, a good beginning would be to replicate some of Piaget's 1920s work on the child's conception of the weather, rivers, mountains, etc. When I first drafted a proposal for the A.A.G. on children's geographic learning in 1974, no research had been carried out on the value of direct experience in children's geographic learning of earth science concepts. "Do children not learn earth science from manipulating, observing and experimenting with mini-landscapes of dirt, sand, mud, water, sticks, stones, etc.?" I asked. "Are not ponds, streams, gutters, gardens, dirt piles, small-scale eco-systems--laboratories for children's dabblings, observations and questions?" I didn't realize that Dennis Wood, also influenced as a student at Clark by the same child advocates as I was--James Blaut, David Stea and William Bunge--was watching his own and his neighbors' children playing with dirt and mud.[50] He seemed a little disappointed that so little of their play involved the modelling of geomorphic or geographic features of any kind, but this was a study conducted over a short period and in only one environment--Raleigh, North Carolina. I have noted that sub-cultures of children have fads which emphasize one kind of play for a whole season, only to change dramatically the

---

[49]. S. Carr and K. Lynch, "Where Learning Happens," *Daedalus: Journal of the American Academy of Arts and Sciences* 97 (1968): 1277-1291.

[50]. D. Wood, "Early Mound-Building: Some Notes on Kids' Dirt Play" (unpublished manuscript, Raleigh, N.C.: School of Design, North Carolina State University, 1976).

next year. For the pre-school children of his study the main focus was cake-making. My own observations, however, revealed a large amount of landscape construction.[51] Hopefully, Wood will continue his research with a larger population in a range of different environments as he has proposed.

Piaget specifically investigated children's understanding of physical systems and processes of the geographic environment in his early books, first published in 1926, 1927 and 1937.[52] A summary of his theory is useful because it remains the most comprehensive, empirically-based theory. He found that children are "both nearer to and farther from the world of objects than we are."[53] They are nearer to it because of a tendency to accept the world as it appears and through a projection of their own feelings into it. At the same time, these tendencies create confusions that distance them from the intellectually apprehended physical work in which adults live most of the time.

For Piaget, the first question is one of "realism," that is, to what extent a child distinguishes the self from the external world. "Realism" should not be confused with "objectivity." The achievement of objectivity involves maximizing one's awareness of the countless intrusions of the self in everyday thought. In contrast, realism means ignoring the existence of the self and thereby regarding one's perspective as absolute. Piaget described three complementary processes involved in the evolution of reality from three to eleven years of age. The first is the progressive differentiation of self from the external world, towards objectivity. Objectivity is never completely achieved; there always remain "adherences," less differentiated fragments of experience. The following five adherences are actually inseparable from one another. "Participation" is the feeling in a very young child that the world is filled with intentions which merge with one's own, e.g., the sun and the moon follow as we walk and the clouds and wind notice us and obey us. "Animism," related to "participation," is the notion that things such as clouds have consciousness. "Artificialism" is the tendency of a person to think that everything is made for and by people. With "finalism" things simply end without their origins or consequences being noticed, and without being necessarily endowed with consciousness, e.g., a river flows so as to go into a lake. "Force" is the idea that things make efforts

---

51. Hart, op. cit.
52. Piaget, 1954, 1960, 1967, op. cit.
53. Piaget, 1954, op. cit., p. 254.

through some kind of energy equivalent to a person's muscular force. With these adherences is "phenomenism," the idea that anything can produce anything: two facts observed together can be causally related.

With the growing objectivity that develops as a child frees the self from these adherences, there is a growing reciprocity with the perspective of others. Gradually a child realizes that his or her own perspective is unique; the classic example is the child who thinks the sun and moon are small globes following us just above roof level as we walk along the street!

The third process, from realism to relativity, is closely related to the achievement of reciprocity, described above. For young children everything has absolute substance and quality. Only gradually do they see phenomena as dependent upon each other and relative to us. They progressively realize that external forces determine the motion of natural objects and that these in turn are dependent upon other forces, forming together a universe of relations. Together with this growing awareness of the relativity of physical phenomena a child has a growing conception that his own ideas are relative to himself and to his personal evaluations of things.

In the many years since Piaget wrote these books there have been many criticisms. A central criticism is that what Piaget takes to be universal developmental uniformities from his research with Swiss children should not be considered laws of nature for they are historically and socially determined.[54] For this reason, one should expect animism, for example, to appear or disappear in the same way in all cultures, for the dominant adult cultures themselves have different conceptions of their physical universe. Nevertheless, the kinds of misinterpretations of reality described by Piaget have been repeatedly found in children.

A small study by Kates and Katz,[55] geographers of Clark University, serves as a classic example of the rich data that can be gleaned by simply observing and listening to children, recording their spontaneous discussions and questions and then discussing with them their thoughts about a phenomena. In this case the subject was the hydrologic cycle as understood by four, five and six-year-old children attending a day care center in Worcester, Massachusetts. One of the authors had spent more than a decade investigating the

---

54. L. S. Vygotsky, *Mind and Society: The Development of Higher Psychological Processes* (Cambridge, Mass.: Harvard University Press, 1978).

55. R. W. Kates and C. Katz, "The Hydrologic Cycle and the Wisdom of the Child," *Geographical Review* 67 (January 1977).

adult world of water resources. Neither of them had experience in working with children, but they learned that children had such interest in the phenomena and talked so freely that it was easy and enjoyable to produce interesting results. Their work confirmed the sequences Piaget outlined on the understanding of physical systems, as described above, but with some interesting differences worthy of further study. The Worcester children of only four years of age had explanations which were less artificialist and more naturalist than those of Piaget, more similar to the six to nine-year-old Swiss children. Whether these differences are due to the different environments of the two groups, the changing access to environments, or information through the media over the fifty year period between the two studies, etc., is entirely open to speculation at this time. Another interesting finding was that the children commonly constructed two separate unlinked cycles: a domestic water cycle and a cloud-rain or natural cycle. No doubt they had difficulty linking the two systems because the natural hydrologic cycle is impossible to observe due to the large scale, slowness and, at times, invisibility of the process. Such understanding, then, can clearly be influenced not only by the intellectual capacity of a child but by the information available. There is much room here for both pure research and interventional programs of education and research.

This research tells us that it is time for us to stop simply replicating Piaget's experiments to see by what age children can understand something. We need to open our research designs to be able to reveal the role of experience. I do not mean experience as reduced to the socio-economic class a child is from, but whether they live in a village or a city, how much they have travelled by car or by air, what access to visual media they have had, etc. In 1934 Lucy Sprague Mitchell wrote:

> The kind of geography which we have found most natural to city children is human geography. The natural earth conditions are too overlaid with human modifications to make physical geography an easy field of exploration in the city. In the country, even in the suburbs, children discover relationships which concern soil, erosion, elevation, the growth of plants and animals, at an earlier age than do city children. But for children before eight everywhere, work, the activities of people, hold more interest than natural phenomena.56

No matter whether one agrees or disagrees with this statement, it is embarrassingly clear that it is at least an informed speculation. Moreover, it is disturbing that neither geographers nor psychologists have any empirical bases for confirming or denying it.

---

56. Sprague-Mitchell, op. cit., p. 24.

Human Geography

There has been, on the surface at least, great interest in teaching young children what is popularly termed "ecology," but there seems to have been very little research on the development of children's interest in, and conception of, people's use of the environment. I observed informally during my own study of children's place experience in a New England town, however, that children were extremely interested in recreating in their toy and dirt play on the ground the occupations of adults in their communities. I have observed two brothers, four and nine years of age, for example, spend hours playing at highway building in the dirt; the older boy recreating all that he has observed in passing workmen on the roads--first laying a heavy course with stones and then layering it with dirt before making a smooth layer by mixing fine dirt with water to make a slushy, shiny tarmac; building retaining walls with large stones at the bottom leading up to fine gravel at the top; finishing up by simulating a thunderstorm with a hose pipe spray to see how well his system holds up under "real" environmental conditions.

On another occasion, while observing a small army of children, all under ten years of age, playing seriously on a dirt/clay hillside, I asked a three-year-old what he was building. "The emergency dam," he told me indignantly as though anyone should know. He went on to explain how his channel and dam were designed to take the water should one of the upper dams, being built by other children, break. No adults had talked with him or any of the children about the complicated system they were building. This kind of learning is not difficult to observe. If we engaged in this type of observation closely and systematically, we would be in a much better position to build theory and to pose better questions to children both in our further research and in our educational practice. Simply to observe the geographic content of children's spontaneous modelling in block corners, sand boxes, sidewalks, bedroom floors, etc., and to see if, and how, this varies according to the children's home environment and parents' occupations, etc., might be a valuable beginning.

In the United Kingdom, a small number of geographers, planners and educators have achieved a remarkably good integration of research with education and have shared this with each other, and with a large number of geography teachers through the *Bulletin of Environmental Education*, published until very recently by the Town and Country Planning Association.[57] Lucy Sprague Mitchell would be thrilled to see dozens of essays outlining ways that teachers can build classes around children's reflections on their own environmental perception and behavior, and through research with residents of their own local communities. The original inspiration for this comes from Patrick Geddes via Colin Ward, original co-editor of *B.E.E.* with Tony Fyson.[58] The content of the articles, published almost every month for the past decade, clearly reflects the growth of behavioral geography. Some of the writers, such as Brian Goodey[59] are geographers, while other regular writers such as Jeff Bishop[60] have been directly influenced by research in behavioral geography. The work goes beyond what Sprague-Mitchell called for, carrying children into an understanding of the social, economic and political forces which influence people's relatedness to the environment and how environmental decisions are made which affect them and their neighborhoods. With this kind of participatory work *with* children, research and practice become one and the same thing. There are excellent opportunities in the U.S.A. to collect data while simultaneously providing enjoyable experiences for children and teachers because the nation's schools are starved for geographic methods and content. For example, a recent project of ours involved thirty schools in four states in which classes of children were twinned with classes with children from very different environments. They corresponded for an entire school year in order to compare their own local environments, what they are like and how they are changing.[61] The project was very successful for the children and teachers, and the letters, drawings, models and interviews are providing us with valuable insights into how rural, suburban and city children perceive their own environments and stereotype others.

---

57. *B.E.E.* has recently been renamed *Streetwork*. It is still available ten times a year, published by Streetwork, Notting Dale Urban Studies Center, 178 Freston Road, London, England.

58. For a summary book on the approach, see C. Ward and A. Fyson, *Streetworks* (London: Routledge and Kegan Paul, 1973); see also C. Ward, *The Child in the City* (New York: Pantheon, 1978).

59. B. Goodey, *A Child's Eye View of Smallheath* (Birmingham, England: Center for Urban and Regional Studies, University of Birmingham, 1972).

60. J. Bishop, "Planners, Children and Teachers," *Bulletin of Environmental Education* 67 (November 1976): 4-11.

61. R. Hart and C. Perez, 1980.

Cognitive Mapping

The largest amount of research in "developmental geography" has focused on children's understanding of the spatial location of phenomena in the landscape. With the genesis of behvioral geography in the later 1960s, it was too tempting to ignore the ready-made term "cognitive map"[62] and to adopt this term as a symbol of the profession's entry into the *terrae incognitae* of the mind.[63] This quickly led to the conceptual misunderstanding that representations of the geographic environment were necessarily map-like and singular. Attempts to clarify the term or extinguish it are periodically made,[64] but still there is a strong tendency, even in the research, to think of the cartesian map as *the* model for our mental representations.

In 1968 Professors Blaut and Stea founded the "Place Perception Project" at Clark University.[65] The Place Perception Project prepared an agenda for research in children's geographic learning, generated new theory, and conducted some provocative research on the early map language abilities of children. The observation that children learn mapping skills before they enter schools, and that, given the right media, teachers can tap into this ability, serves as a fine demonstration that geographic learning happens spontaneously and that the schoolteacher's job is to be aware of this and to assist the process.

Without doubt the theory and research by Piaget and his colleagues on spatial cognition has had the greatest influence on research in spatial cognition of the geographic environment.[66]

---

62. E. C. Tolman, "Cognitive Maps in Rats and Men," *Psychological Review* 55 (1948): 189-208, extracted in *Cognitive Mapping: Images of Spatial Environments*, ed. R. Downs and D. Stea (Chicago: Aldine-Atherton, 1973).

63. R. Downs and D. Stea, eds., *Cognitive Mapping: Images of Spatial Environments* (Chicago: Aldine-Atherton, 1973), for a history and summary.

64. E. G. Hart and Moore, op. cit.; Downs and Stea, op. cit.; R. M. Downs, "Maps and Mappings as Metaphors for Spatial Representation," in Liben et al., op. cit.

65. J. Blaut, *Studies in Environmental Geography,* Place Perception Research Report no. 1 (Worcester, Mass.: Graduate School of Geography, Clark University, 1969); J. M. Blaut and P. Stea, *Place Learning,* Place Perception Research Report no. 4, (Worcester, Mass.: Graduate School of Geography, Clark University, 1969); J. M. Blaut and D. Stea, "Studies of Geographic Learning," *Annals of the Association of American Geographers* 61 (1971): 387-393; J.M. Blaut, G. McCleary and A. Blaut, "Environmental Mapping in Young Children," *Environment and Behavior* 2 (1970): 335-349; D. Stea and J. M. Blaut, "Notes toward a Developmental Theory of Spatial Learning," *Proceedings of the Second Conference of the Environmental Design Research Association* (Pittsburgh: Carnegie Mellon University, 1970); D. Stea and J. M. Blaut, "Some Preliminary Observations of Spatial Learning in School Children," in Downs and Stea, op. cit., 1973.

66. Reviews by Hart and Moore, op. cit.; Siegel and White, op. cit.; Hart, 1978, op. cit; G. T. Moore, "Theory and Research in the Development of Environmental Knowing," in *Environmental Knowing,* ed. G. T. Moore and R. G. Golledge (Stroudsberg, Penn.: Dowden, Hutchison & Ross, 1976); D. Piche, "Spontaneous Geography of the Urban Child," in *Geography and the Urban Environment,* vol. 4, ed. D. Herbert and R. G. Johnston (New York: Wiley, 1982).

Piaget has amply demonstrated through a large number of experiments that children actively construct their spatial knowledge.

According to Piaget, children build schemas, sequences of behavior which are internalized in thought to become the basic building blocks of cognition. Intelligence is internalized action. He outlined four major periods of development: the *sensori-motor* period from birth to approximately two years of age, during which an infant proceeds from reflex activity only, to coordinated actions in space. A child begins to internalize these behavioral schema and to enter into the *pre-operational period* at about two years of age. Now the actions can be carried out symbolically but these mental operations on the spatial schema are only intuitive. For example, a child can reverse thinking only by starting again at the beginning of any sequence--a cyclical rather than a true reversability with obvious implications for the learning of routes in geographic space. Thinking is egocentric, that is, a child has difficulty decentering from any one aspect of a situation. With the *concrete operational period* beginning around seven years of age, a child becomes capable of logical thought. For example, it now becomes possible to achieve true reversibility of thought, though these operations are still limited to real objects. The child no longer confuses his own point of view with that of others; different points of view, independent of the self, can be differentiated and coordinated into a comprehensive knowledge of spatial relationships. Entrance into the final phase identified by Piaget, the *formal operational period,* during adolescence, means that a child is no longer dependent upon the manipulation of real objects, it is now possible to operate strictly with ideas with some kind of language such as words, diagrams or abstract maps. This sequence of development has been investigated and challenged in hundreds of experiments by investigators around the world. In the U.S.A. much attention has been wasted upon arguments about how early a stage is reached and how this relates directly to teaching[67] In spite of the arguments, the major principles of Piaget's theory are now well accepted. The most important of these for behavioral geography is the clear demonstration that even our adult understandings and representations of space result from the extensive manipulation of objects and from locomotion through environments, not as we might intuitively feel, from our "reading" of the environment.

---

67. See critique by Duckworth, 1982.

There have been a large number of studies in the past decade, primarily based on Piaget's theory, on the development of children's representation of geographic space.68 Some of this research has modified Piaget's experiments specifically for their relevance to geographic education.69 Most, however, has been conducted from within developmental psychology where it has grown to become a significant sub-field.70 Unfortunately, within this discipline, research suffers from a heavy reliance upon experimental simulation of the geographic environment and has made little attempt to build research upon the very different goals people have when they mentally represent places.71 There are methodological reasons why cognitive developmental psychologists have not conducted more naturalistic research in the large-scale environment.72 Many, including Piaget and his colleagues, argue that the best approach to cognitive development is to look at a child in the process of problem-solving, that is, while manipulating the phenomena. It is, of course very difficult to do this with the geographic scale environment. The solution has been to simulate the geographic environment through the design of model settings, mazes, and the use of blindfolds. Consequently, the geographic environment has been reduced in scale and simplified to the point where the validity of the research suffers greatly. Research could benefit from more collaboration with geographers and environmental psychologists who are more experienced with field research designs and methods. In spite of this weakness, the research has managed to develop a network of researchers who are steadily moving forward through the development and testing of theory with some ingenious simulations of geographic spaces.73 The major debate between the incrementalists

---

68. See summaries and reviews by Hart and Moore, op. cit; Siegel and White, op. cit.; G. Moore, 1976, op. cit.; Hart, 1978, op. cit.; Liben et al., op. cit.; L. P. Acredolo, "New Directions for Environmental Cognition: We Can Get There from Here," paper presented at the Annual Meeting of the Eastern Psychological Association, New York, 1976 (available from the author, Department of Psychology, University of California at Davis).

69. J. Towler, "The Dissemination of Research in Geographic Education," *Journal of Geography* 71 (1972): 238-240; J. Towler, "The Elementary School Child's Concept of Reference Systems," *Journal of Geography* 69 (1970): 89-93.

70. There were three separate sessions on spatial representation and way-finding in the last conference of the Society for Research in Child Development (1981).

71. R. Hart, "Children's Spatial Representation of the Landscape: Lessons and Questions from Field Study," in Liben et al., op. cit.; R. Hart and M. Berzok, "Children's Strategies for Mapping the Geographic-Scale Environment," in *Spatial Abilities: Development and Physiological Foundations,* ed. M. Potegal (New York: Academic Press, 1982).

72. Discussed by Piche, op. cit.

73. L. P. Acredolo, "Frames of Reference Used by Children for Orientation in Unfamiliar Spaces" (Ph.D. dissertation, Institute for Child Development, University of Minnesota, 1974); M. Gauvain, "Children's Exploration and Memory of Spatial Routes and Layouts" (Ph.D. dissertation, Department of Psychology,

and the constructivists continues. They disagree as to whether spatial cognition is largely the result of the degree of experience with an environment, or whether it is due to the intellectual abilities which are applied to it.[74]

While research by developmental psychologists has completely ignored the important "content" side of spatial representation, some of the work by geographers and planners described under "The Geography of Children" above, included "cognitive mapping" and looked simultaneously at *what* is represented as well as *how* phenomena are related one to another.[75] Developmental psychologists wishing to broaden their perspective on children's spatial cognition would do well to look outside of their field to those more ecologically valid kinds of descriptions of the child's spatial world. One important discovery they would make is that way-finding is not the only or even the primary reason that children represent to themselves the spatial properties of the large-scale environment. Among other reasons, we may hypothesize that a child learns about the geographic surroundings because they are intrinsically interesting.

The work of Tindal[76] and Anderson and Tindal,[77] described under "The Geography of Children," on the spatial behavior and neighborhood mapping ability of children from different socio-economic backgrounds and ages was a valuable beginning to research on the influence of social, cultural, and environmental background on geographic experience and spatial knowledge. In a similar vein is the work of another geographer on the "home range and urban knowledge" of children.[78] As dissertation research, Michael Southworth with teenagers, and I with younger children and pre-adolescents, investigated simultaneously the relationship between spatial behavior and knowledge of the spatial properties of the environment.[79] This work now needs to go beyond the case study approach through the development of comparative methods.

University of Utah, 1982); H. L. Pick, Jr., and J. J. Lockman, "The Development of Spatial Cognition in Children," in *Mind, Child and Architecture,* ed. J. C. Baird and A. D. Lutkus (Hanover, MA: University Press of New England, 1982); A. Siegel, "Methodological and Meta-Theoretical Issues in the Development of Cognitive Mapping of Large-Scale Environments," in Baird and Latkus, op. cit.

74. See arguments by Pick and Lockman, op. cit.; Siegel, op. cit.; review by Piche, 1982.

75. E. G. Southworth, op. cit.; Lynch, op. cit.; Hart, 1978, op. cit.; Moore and Young, op. cit.; R. Moore, op. cit.

76. Tindal, op. cit.

77. Anderson and Tindal, op. cit.

78. H. Andrews, "Home Range and Urban Knowledge of School-Age Children," *Environment and Behavior* 5 (1973): 73-86.

79. Southworth, op. cit.; Hart, 1978, op. cit.

Geographic Cognition Beyond the Local Environment

There has been much less investigation of children's spatial knowledge beyond the home environment. What conceptions do children have of places beyond the horizon? How do they organize different places and regions? What role does the imagination play in these developments? Very little research or even theoretical speculation has been carried out of these questions. Peter Gould, a geographer, asked thousands of Swedish school children what place names they knew.[80] This might be an interesting area to investigate, but to do so one needs to be much more concerned with conceptual issues at such an exploratory stage of research. After manipulating the large amount of data, Gould found that more children knew more place names of larger places than of smaller places and that this place-naming ability is neatly predicted by a gravity model. This may seem elegant but it tells us little about children and their geographical worlds. We need to work with some of the conceptual expertise psychology has developed. Other information about the Swedish children than simply the location of their homes, for example, might have helped in providing some useful interpretation of the data.

Piaget and others investigated the development of children's concept of nation but this work has fitted into the rather narrow question of the development of class inclusion abilities in children.[81] Piaget[82] found when he asked children if they were Genevese or Swiss that "pre-operational children" (i.e., approximately 7 years of age or younger) responded in terms of "nominal realism," seeing Switzerland as a separate place alongside the one where they lived. During the "concrete operational stage" (from approximately 7 to 12 years of age) they visualized Switzerland as an area surrounding their city and only with the achievement of formal logical abilities, in Piaget's final stage (12 years and over), were children able to conceive of city and nation

---

80. P. R. Gould, "The Black Boxes of Jonkoping: Spatial Information and Preference," in *Image and Environment: Cognitive Mapping and Spatial Behavior*, ed. R. M. Downs and D. Stea (Chicago: Aldine, 1973); P. Gould and P. White, *Mental Maps* (Harmondsworth, Middlesex, England: Penguin, 1973).

81. J. Piaget, *Judgement and Reasoning in the Child* (London: Routledge and Kegan Paul, 1926, French edition, 1924); J. Piaget and A. Weil, "The Development of Children of the Idea of the Homeland and Relations with Other Countries," *International Social Science Bulletin* 3 (1951): 561-578; G. Jahoda, "The Development of Children's Ideas about Country and Nationality, Part 1: The Conceptual Framework," *British Journal of Educational Psychology* 33 (1963): 47-60; G. Jahoda, "The Development of Children's Ideas about Country and Nationality, Part II: National Symbols and Themes," *British Journal of Educational Psychology* 33 (1963): 143-153.

82. Piaget, 1924, op. cit.

as part of a geographic hierarchy. Jahoda[83] conducted similar research with children aged 6 to 11, asking them about the relationship of towns and counties based on their home in Glasgow. The children's verbal answers could not all be fitted into Piaget's description of stages. He argued that children's understanding of spatial relations does not necessarily precede their understanding of nationality. When he gave them abstract two-dimensional forms to express the spatial relationship of nations their performance was inferior, suggesting that they were not fully understanding what they were verbally expressing. Jahoda had sample populations drawn from working class and middle class sectors of Glasgow.[84] Here was a wonderful opportunity to investigate the influence of experience on geographic knowledge by simultaneously investigating differences in children's travel mobility, and access to media, etc. Instead, Jahoda simply tells us that working class children performed worse on the tasks, leaving the reader to assume what they wish!

An example of what can happen when psychology and geography integrate can be seen in the excellent recent geography dissertation on "children's spontaneous geography" by Denise Piche.[85] Piche went beyond the study of children's mental home range to include their conception of distant places and regions. This research was conducted more in the spirit of the constructivist investigator Piaget than has been true of other research in children's cognitive mapping. Rather than focussing upon the products made by children—drawings, maps, or models—she attended to the *process* of making them. Using a clinical interview approach, she asked them about places in their neighborhood and "elsewhere" and then proceeded with different questions for each child but always with the same intention of examining

> how each individual divided the continuity of space into places, interpreted the names of places and land-uses and structured the continuity of geographic space, including how they hierarchically organized geographic concepts and finally explained, identified with, and judged the world.[86]

Piche's findings confirmed the main sequence of development outlined by Piaget, but her findings offer a clear call to go beyond the almost total emphasis of current research upon the proximate environment. She argues that we should simultaneously investigate children's understanding of proximate and distant space for "a

---

83. Jahoda, 1963, 1964, op. cit.

84. See also J. P. Stoltman, "Children's Conception of Territory: A Study of Piaget's Spatial Stages" (Ph.D. dissertation, University of Georgia, 1971).

85. Piche, op. cit.

86. Ibid., p. 234.

correct conceptualization of here is necessarily related to a conception of 'elsewhere'." It is regrettable that more details of her analysis cannot be presented here for it reveals clearly how many very interesting questions remain to be explored on children's "spontaneous geography" beyond the home environment.

It is worthwhile pausing to ask why there has been such a narrow emphasis upon children's cognitive mapping when there are so many other important domains of geographic cognition to investigate. There are numerous possible explanations. One likely reason is the attractiveness of maps themselves. As Roger Downs has recently argued, it has been extremely difficult to convince investigators in the field that the map is only a visual metaphor for whatever our mental representations may be like.[87] The map is the one tool and symbol geography has been able to hang onto as distinctly belonging to its discipline. It is a small wonder that children's geographic research should begin with it. It would be good to be able to say that the spatial emphasis is the result of an applied orientation in the research such as map skill curriculum development, spatial orientation schemes, signing or public information for children. This is not the case. The major reason, I believe, is the result of the tendency of academia to establish sub-fields and blindly produce studies without asking each time, "what are the important questions to ask about children's geographic knowledge?" Closely related to this is a kind of methodological determinism: if you have a good spatial cognition method, use it. Again, the love of maps is a good example--there probably is no introductory behavioral geography class of students that has not been asked to draw a "mental map," usually without much critical attention to the meaning of such a product or how one goes about analyzing it. It is time for those of us in this sub-field of behavioral geography to return to the pioneering study, *The Image of the City* by Lynch[88] and read the Appendix. Here one finds a rich description of the different possible reasons why people, including children, might want to mentally represent the spatial properties of the environment. There are many reasons and they do not all call for the same kind of cartesian map; in fact, it may be too circumscribed an approach to begin with the assumption that the primary image of a place is map-like at all. Shelley Pazer, a developmental psychologist at the City University of New York, is intrigued by some of the special questions children's geographic cognition brings to her field of

---

87. Downs, 1981, op. cit.
88. K. Lynch, *The Image of the City* (Cambridge, MA: MIT Press, 1960).

children's concept formation.[89] She is investigating the development of children's concept of "city." Such phenomena raise exciting new questions for developmental psychologists. Rather than beginning with spatial organization of the city as *the* question and hence maps as *the* method, her research design remains open in order to reveal the relative roles of form versus function, an important debate in the literature of children's concept formation.[90]

The Place of Geography in Future Research with Children
Other "Geographies of Children" and "Children's Geographies"

The research to date is better characterized by what it has not studied than through the small number of reports currently published. We know very little, for example, of children's social geography, commercial geography, natural resource knowledge and use, spatial diffusion of games and other cultural "inventions," the influence of early spatial experience or environmental experience on adult behavior, or children's changing conceptions of the universe, to name just a few.

Bridging the Gap Between Micro and Macro Research

There is a need for research which is "ecological" in Bronfenbrenner's sense of the term.[91] Bronfenbrenner argues that for our research of children's lives to be ecological we must go beyond the naive interpretation of Kurt Lewin[92] as a need to study a child in his or her total immediate context. Bronfenbrenner emphasizes the need to conceptualize research in such a way that different levels of analysis and different settings can be linked to one another. Micro level ecological studies such as children in playgrounds, for example, can be linked with studies in other settings like the home by doing research on the relationship of behavior in one setting with behavior in another setting. He argues further, calling for study of the influence on non-child settings in which child-caretakers spend their time, notably the workplace, on the quality of their relation to children in children's micro-settings. By studying the relationship of these institutions,

---

89. S. Pazer, "Children's Conception of the City" (dissertation proposal, Developmental Psychology Program, City University of New York, 1983).

90. See also J. Muntanola, *L'Apprender la Ciudad* (unpublished working paper, School of Architecture, University of Barcelona, 1982).

91. U. Bronfenbrenner, "Lewinian Space and Ecological Substance", *Journal of Social Issues* 33 (1977): 198-213; U. Bronfenbrenner, *The Ecology of Human Development*: *Experiments by Nature and Design* (Cambridge, MA: Harvard University Press, 1979).

92. K. Lewin, "Behavior and Development as a Function of the Total Situation," in *Field Theory in Social Science,* ed. K. Lewin (New York: Harper, 1951, originally published 1946).

normally investigated separately, he believes we would be able to plan for, and assess the impact of new national or state level policies and institutional arrangements. While such a scheme sounds more feasible for policy-making in a socialist government than in the United States, this kind of thinking is refreshing from a psychologist. The role geography could contribute to such an ecological research endeavor is great. A particularly valuable development would be for behavioral geography with its concern for individual's experience of the environment, to join with those geographers who limit their research to the macro-economic/political analysis of social justice. Such a combination could lead to the kind of research on the geography of children called for by Bunge in which the forces behind variations in the availability, accessibility, and quality of resources for children are identified. Then, we would be in a position to make specific policy and planning recommendations for those children and families who most need our geographic expertise.

Geographic Learning Research for Geographic Education

Unfortunately, few educators or teachers follow the excellent lead of Lucy Sprague-Mitchell[93] who built her educational program out of the geographic interests and activities of children in her classrooms and those of her students and colleagues. Ideally, teachers would themselves each become investigators, building their teaching upon observations of their children and their spontaneous questions. There is also much room, however, for basic research contributions. This geographic learning research should not be harnessed to specific curriculum design questions; it should rather focus upon the questions children themselves have and how they proceed to find answers to them. This research would best be done collaboratively between geographers and developmental psychologists, working with educators to help interpret the findings for their relevance to educational practice. We must avoid the kind of isolated research endeavour which characterizes the current spatial cognition literature.

Research on the Geography of Children for Environmental Planning and Design

It is clear from the existing research reviewed that children have different land-use categories from those of adults. If we could develop a reliable comparative method it would be very useful in helping landscape planners and designers, particularly in the

---

93. Sprague-Mitchell, op. cit.

creation of new residential settings for children. They are
currently operating with only the adult glossary of land-uses when
the major users of outdoor open spaces are children. Recent
attempts by Lynch[94] and Moore[95] with small comparative samples are a
beginning. We now need to concentrate on validating some truly
comparative methods. Our results would not be cookbooks for the
provision of landscape resources for children, for this can not be a
value-free enterprise. The research would, however, provide
valuable insights for planners and designers of the very different
scale and qualities of features valued by children. For the actual
planning process, simpler participatory research would be ideal,
particularly in the improvement of existing residential areas. The
behavioral geographers would then find themselves in the dual roles
of informed consultants summarizing what we know from existing
research with children, and research coordinators establishing
simple indices of children's existing spatial access to resources
and obtaining land-use priorities from both children and parents.
This would guarantee sensitivity to the particular culture or
sub-culture and to the local environment. Models for this kind of
learning process are hard to find though the *British Bulletin of
Environmental Education* has many valuable examples of ways to help
children planning and design.

The Physical Environment as an Agent in Cultural Reproduction

We need to build theory on human development through the
environment not only for its value in understanding children, but
also because of the longer term question of cultural development in
relation to the environment. Developmental psychology has focussed
almost entirely upon the isolated investigation of physical, social
and abstract cognitive development, leaving us with very little
sense of how children's daily transactions with a dramatically
changing environment are themselves changing and what the
implications of these changes might be for children and for the
future cultures they are in the process of creating.

American psychology has been distinctly non-materialistic and
ahistorical. There are strong signs by many in developmental
psychology of a desire to become more contextual. Particularly
influential have been the recent translations of the writings of
Soviet psychologists Vygotsky[96] and Luria.[97] Cultural and behavioral

---

94. Lynch, 1977, op. cit.
95. R. Moore, 1981, op. cit.
96. Vygotsky, op. cit.

geographers are in a good position to help developmental psychology move into field research, not necessarily dropping its concern with experimental research design, but leaving the "no-place" world of laboratories behind in recognition that children's engagement with the material world varies in ways which may have important consequences for their lives. Fifty years ago Luria set out, under Vygotsky's influence, to investigate the impact of different degrees of contact with technology and collectivization on peasant's thinking in Soviet Central Asia. By interviewing a cross-section of peasants with dramatically different degrees of removedness from the advancing Soviet culture, Luria was able to investigate the impact of historical changes in technology on people's consciousness. A related research design has been chosen by Cindi Katz, a geographer at Clark University, to investigate the impact of a changing economy and associated technological changes on children's environmental learning.[98] A short time ago she returned from a village in the Sudan which has recently been transformed from a subsistence agricultural economy to a market-crop economy. Through ethnographic and ethnosemantic procedures she reconstructed with a small number of children an account of their environmental knowledge. The research combines the best field research traditions of cultural anthropology, microcultural geography, and behavioral geography with a Marxist analysis of major structural changes which it is hypothesized are impacting individual behavior and environmental learning. I hope the work of Katz will lead cultural geographers to realize that the study of children's environmental learning offers valuable access to the understanding of people's environmental knowledge and values as well as to the larger debate on the relationship between environmental practice, culture and consciousness.

---

97. A. R. Luria, *Cognitive Development: Its Cultural and Social Foundations,* trans. M. Cole (Cambridge, Mass: Harvard University Press, 1976).

98. C. Katz, "Children's Environmental Learning and Knowledge in a Changing Social Context in Riverine Arab Sudan" (Ph.D. proposal, Graduate School of Geography, Clark University, 1979).

CHAPTER 8
ENVIRONMENTAL PERCEPTION, HISTORIC PRESERVATION, AND SENSE OF PLACE

Robin E. Datel and Dennis J. Dingemans

The impulse to preserve valued vestiges of the past is powerful in modern society. Geographers increasingly are interested in that impulse as it makes a growing impact on the human landscape.[1] They have begun to study the many activities, from public regulation to private restoration, that comprise historic preservation. The first part of this paper reviews briefly those studies. It is our contention that environmental perception approaches are especially appropriate, since many questions about historic preservation concern the "world in the mind." Next, our paper discusses how this initial body of work can be expanded through research efforts that explicitly consider historic preservation as part of the search for a sense of place. We believe that both scholars and practitioners would benefit from continued refinement of environmental perception methodologies of the type required for research on sense of place. Lastly, the paper comments on comparative studies of historic preservation that address perception issues and reports some findings of our own.

Environmental Perception Studies of Historic Preservation

The logical match between historic preservation and geography is becoming better and better as both fields evolve. Preservation is shifting its attention from "aesthetic and historical significance to the community value of structures and landscapes" and to things "familiar and well-loved."[2] At the same time, geographers are becoming more sophisticated in their work on sense of place and on attitudes toward the visible past. Despite this

---

1. This growing impact is described in Donald Appleyard. ed., *The Conservation of European Cities* (Cambridge, Mass.: MIT Press, 1979); Nathan Weinberg, *Preservation in American Towns and Cities* (Boulder, Colo.: Westview Press, 1979); and James Marston Fitch, *Historic Preservation* (New York: McGraw-Hill, 1982).

2. David Lowenthal, "Environmental Perception: Preserving the Past," *Progess in Human Geography* 3 (September 1979): 555.

growing goodness of fit between historic preservation and geography, a recent review of the literature by David Lowenthal was somewhat critical.[3] It pointed out lost opportunities and the small number of research contributions. Only a handful of geographers make historic preservation their research specialty, and their work is not well-known, compared to the research of historians, architects, and planners.

Geographers studying historic preservation have stressed the measurement of real-world phenomena rather than preferences, attitudes, or experiences. Typical work explores the socio-economic impacts of preservation, such as reversing inner-city decline, gentrifying neighborhoods, or bringing the economic benefits of recreation to historical resource areas. For example, Ford reviewed the role of historic preservation in changing the physical and social character of American towns, urban cores, zones of discard, and older neighborhoods.[4] The search for a sense of place is acknowledged as an important cause of the upsurge of preservation activity, and Ford argues that "models of city structure must consider values and changing images of the ideal as well as such things as rent gradients and traffic patterns."[5] But studies that focus on those values and ideals--with particular reference to historical features--in Charleston, Chicago, Seattle, Columbus, Boston, or St. Louis (the cities he used as examples) or other cities are largely absent. Lowenthal suggests that the emphasis on socio-economic and recreation management aspects of historic preservation is to be expected because "geographers in general, and Americans in particular, adopt an unsentimental, no-nonsense, hard-headed approach to historic preservation."[6]

Datel and Dingemans recently reviewed the geographical literature that addresses the questions of why we preserve, how we preserve, and where we preserve.[7] They identified two main clusters of work on preservation that address aspects of environmental perception. The first used primarily literary and historical sources to describe national or universal attitudes to the past and to preservation. The preeminent scholar in this humanistic mode has been David Lowenthal, who has written about attitudes to the past,

---

3. Lowenthal, "Environmental Perception," pp. 549-559.

4. Larry R. Ford, "Urban Preservation and the Geography of the City in the USA," *Progress in Human Geography* 3 (June 1979): 211-238.

5. Ford, "Urban Preservation," p. 235.

6. Lowenthal, "Environmental Perception," p. 552.

7. Robin E. Datel and Dennis J. Dingemans, "Historic Preservation and Urban Change," *Urban Geography* 1 (1980): 229-53.

historical artifacts, and historic preservation in the United States and with Prince, in Britain.8 How people perceive, experience, and manipulate the past in the landscape is related to broad trends in social, political, religious, and aesthetic thought and experience. The temporal dimension--how attitudes toward the past in the landscape have changed over time--has been discussed in the American context by Lowenthal and in the English one by Prince.9 Jackson has explained the current American restoration mania as deriving from a view of United States history as a Golden Age followed by a Fall.10 Kearns has described how deep-felt resentment of the Anglo-Irish poses a major barrier to the preservation of Georgian Dublin, which is viewed as an Anglo-Irish creation.11 Rowntree and Conkey have interpreted historic preservation in Salzburg, Austria as a stress-induced symbolization process.12 Lowenthal has written in general terms about the nature of human relationships with artifacts of the past.13 Similarly, Tuan's writings on attitudes and behaviors toward the past, which draw on a breadth of evidence from literature, the humanities, and the social sciences, tend to be in the form of universally applicable generalizations, rather than place-specific research results.14 Pierce Lewis and M. R. G. Conzen have discussed the general rationales for historic preservation.15

---

8. David Lowenthal, "The American Way of History," *Columbia University Forum* 9 (summer 1966): 27-32; David Lowenthal, "The American Scene," *Geographical Review* 58 (January 1968): 61-88 (especially pp. 76-81); David Lowenthal, "The Bicentennial Landscape: A Mirror Held Up to the Past," *Geographical Review* 67 (July 1977): 253-67; and David Lowenthal and Hugh C. Prince, "English Landscape Tastes," *Geographical Review* 55 (April 1965): 186-222 (especially pp. 204-221).

9. David Lowenthal, "The Place of the Past in the American Landscape," in *Geographies of the Mind,* ed. David Lowenthal and Martyn J. Bowden (New York: Oxford University Press, 1976): 89-117; Hugh C. Prince, "Revival, Restoration, Preservation: Changing Views about Antique Landscape Features," in *Our Past Before Us: Why Do We Save It?*, ed. David Lowenthal and Marcus Binney (London: Temple Smith, 1981): 33-49.

10. John Brinckerhoff Jackson, *The Necessity for Ruins and Other Topics* (Amherst, Mass.: University of Massachusetts Press, 1980), pp. 88-102.

11. Kevin C. Kearns, "Preservation and Transformation of Georgian London," *Geographical Review* 72 (July 1982): 270-290.

12. Lester B. Rowntree and Margaret W. Conkey, "Symbolism and the Cultural Landscape," *Annals of the Association of American Geographers* 70 (1980): 459-474.

13. David Lowenthal, "Past Time, Present Place: Landscape and Memory," *Geographical Review* 65 (January 1975): 1-36; David Lowenthal, "Age and Artifact: Dilemmas of Appreciation," in *The Interpretation of Ordinary Landscapes,* ed. D. W. Meinig (New York: Oxford University Press, 1979): 103-128.

14. Yi-Fu Tuan, *Space and Place*: *The Perspective of Experience* (Minneapolis: University of Minnesota Press, 1977) (especially chapters 12 and 13); Yi-Fu Tuan, "Rootedness versus Sense of Place," *Landscape* 24: 3-8; and Yi-Fu Tuan, "The Significance of the Artifact," *Geographical Review* 70 (October 1980): 462-72.

15. Peirce E. Lewis, "The Future of the Past: Our Clouded Vision of Historic Preservation," *Pioneer America* 7 (July 1975): 1-20; M. R. G. Conzen, "Geography and Townscape Conservation," *Giessener Geographische Schriften 35* (1975): 95-102.

A second group of researchers, working in a more empiricist mode, have used survey techniques to gather data on attitudes toward historical buildings. Several studies have been concerned with how people value and behave toward particular restoration neighborhoods.[16] What these surveys found about historic preservation often was fairly bald, namely, that historic buildings were very important to the people in these neighborhoods. Among the more interesting findings were those of Ford and Fusch concerning people living outside but near a restoration neighborhood (German Village in Columbus, Ohio). Attitudes were generally positive, despite the possibility of increased rents and taxes. A survey in a very different setting, villages around Aarhus, Denmark, revealed that "a majority of people liked old, half-timbered houses as a landscape amenity at the same time that they expressed a preference to continue living in a particular type of housing they presently occupied, be that modern single-family home or traditional thatched farm cottage."[17] Clearly people value places that they do not choose to live in; presumably it is this situation that makes possible public subsidies to owners of historical properties.

Two surveys trying to uncover personal dispositions toward the past have been done by geographers. Janiskee studied why urbanites liked to visit festivals in rural relic landscape settings.[18] Restfulness, sense of place, and escape from the city were the major dimensions of the festivals' attractiveness. A more comprehensive research project using a similar methodology studied Torontans' psychological dispositions toward the past.[19] Factor analysis of the results of responses to personality assessment tests resulted in the identification of four major orientations toward the past, concerned

---

16. Richard W. Travis, "Place Utility and Social Change in Inner-city Historic Space: A Case Study of German Village, Columbus, Ohio" (Ph.D. dissertation, University of Illinois, 1972); Larry R. Ford and Richard Fusch, "Historic Preservation and the Inner City: The Perception of German Village by Those Just Beyond," *Proceedings of the Association of American Geographers* 9 (1976): 110-14; Larry R. Ford and Richard Fusch, "Neighbors View German Village," *Historic Preservation* 30 (July-September 1978): 37-41; John O'Loughlin and Douglas C. Munski, "Housing Rehabilitation in the Inner City: A Comparison of Two Neighborhoods in New Orleans," *Economic Geography* 55 (January 1979): 52-70; Sallie M. Ives, "A Symbolic Interaction Approach to the Place Meanings in a Historic District: A Case Study of Annapolis, Maryland," (Ph.D. dissertation, University of Illinois, 1977).

17. Robert M. Newcomb, *Planning the Past: Historical Landscape Resources and Recreation* (Folkestone, Kent: Wm Dawson, 1979): 219. See also Robert M. Newcomb, "The Aarhus, Denmark Village Project: Applied Geography in the Service of the Municipality," *Geographical Review* 67 (January 1977): 86-92.

18. Robert L. Janiskee, "City Troubles, the Pastoral Retreat, and Pioneer America: A Rationale for Rescuing the Middle Landscape," *Pioneer America* 8 (January 1976): 1-7.

19. S. Martin Taylor and Victor A. Konrad, "Scaling Dispositions Toward the Past," *Environment and Behavior* 12 (September 1980): 283-307. See also Victor A. Konrad, "Orientations Toward the Past in the Environment of the Present: Retrospect in Metropolitan Toronto," (Ph.D. dissertation, McMaster University, 1970).

with conservation of historical resources, appreciation of the past as cultural heritage, appreciation of direct experience with the past, and general interest in the past. These four dispositions varied with social characteristics and correlated with specific attitudes toward and behavior in historical environments. Further systematic study of the different ways in which people care about the past will help us judge the wider applicability of the findings of this pioneering effort.

Historic Preservation and Sense of Place

In this section we report on what has been written about the relationship between sense of place and historic preservation. We then go on to argue that if sense of place were better understood, its handmaiden, historic preservation, might be employed more sensitively and in a wider range of valued environments. The concept of sense of place has become important in the environmental perception research mode that Saarinen and Sell appropriately have called the "humanist upsurge."[20] Here we use the term "sense of place" to refer to the complex bundle of meanings, symbols, and qualities that a person or group associates (consciously and unconsciously) with a particular locality or region. This bundle derives not only from direct contact with the place, but from many other experiences and from literary and artistic portrayals.[21] Scholars interested in unwrapping sense of place bundles do well to investigate historic preservation, a process whose self-stated goal is maintaining a traditional sense of place. Conversely, those who wish to understand why and what people preserve need to elucidate the sense of place that inspires their subjects.

Many writers see historic preservation as a means of achieving a stronger sense of place. Ford, for example, demonstrated how English conservation efforts help to retain the special character of Bath, Chester, and Norwich.[22] He also reviewed in a paper called "Historic Preservation and the Sense of Place" how historic

---

20. Thomas F. Saarinen and James L. Sell, "Environmental Perception," *Progess in Human Geography* (1980): 525-49; reference is on p. 531.

21. The concepts of "place" and "sense of place" are discussed in Brian Goodey, *Images of Place: Essays on Environmental Perception, Communications and Education*, Occasional Paper no. 30 (Birmingham, England: University of Birmingham Centre for Urban and Regional Studies, 1974); E. Relph, *Place and Placelessness* (London: Pion, 1976); Yi-Fu Tuan, "Space and Place: Humanistic Perspective," *Progress in Geography* 6 (1974): 221-252; Yi-Fu Tuan, "Place: An Experiential Perspective," *Geographical Review* 65 (April 1975): 151-165; and in Tuan, *Space and Place*. A number of the essays in Anne Buttimer and David Seamon, eds., *The Human Experience of Space and Place* (New York: St. Martin's Press, 1980) and in David Ley and Marwyn S. Samuels, eds., *Humanistic Geography: Prospects and Problems* (Chicago: Maaroufa Press, 1978) also are relevant.

22. Larry R. Ford, "Continuity and Change in Historic Cities: Bath, Chester, and Norwich," *Geographical Review* 68 (July 1978): 253-273.

preservation has salvaged distinctive remnants of the past in many American downtowns being homogenized by urban renewal.[23] In a similar vein, Lewis in a paper entitled "To Revive Urban Downtowns, Show Respect for the Spirit of the Place," argued that we should use historic preservation to retain the *genius loci* of American cities.[24] Yi-Fu Tuan in *Space and Place* reminded us that the pamphlet literature and public rites of Beacon Hill in Boston, which celebrate the architectural beauty, historical events, and notable people associated with the neighborhood, contribute to its rich sense of place.[25] More generally, "awareness of the past is an important element in the love of place."[26] Kevin Lynch has suggested that certain kinds of historic preservation be used to enhance people's senses of place.[27]

Relph, in contrast to the above examples, has written that historic preservation can be antithetical to the development of a sense of place. That conclusion is based upon equating historic preservation with "museumisation," a species of "disneyfication" that creates "absurd, synthetic places made up of a surrealistic combination of history, myth, reality and fantasy that have little relationship with a particular geographic setting."[28] Such places as reconstituted pioneer villages, restored castles, and reconstructed forts are considered to be the result of inauthentic attitudes to place and of place-making according to the dictates of *kitsch* and *technique*. Lewis, despite his argument that historic preservation can help save what is special about a place, has warned of the potential costs of successful historic preservation programs. In New Orleans' French Quarter, the physical legacy of the past was saved, but other elements of the district's traditional sense of place, relating to function and social character, were destroyed.[29] Tuan has noted the limits to what historic preservation can achieve. It can help us enhance our sense of place, which depends upon cultivating a consciousness of the differences between places and

---

23. Larry R. Ford, "Historic Preservation and the Sense of Place," *Growth and Change* 5 (April 1974): 33-37.

24. Pierce Lewis, "To Revive Urban Downtowns, Show Respect for the Spirit of the Place," *Smithsonian* 6 (September 1975): 33-40.

25. Tuan, *Space and Place,* pp. 171-72.

26. Yi-Fu Tuan, *Topophilia*: *A Study of Values* (Englewood Cliffs, N.J.: Prentice-Hall): 99.

27. Kevin Lynch, *What Time Is This Place?* (Cambridge, Mass.: MIT Press, 1972); Kevin Lynch, *Managing the Sense of a Region* (Cambridge, Mass.: MIT Press, 1976).

28. Relph, *Place and Placelessness,* p. 95.

29. Lewis, "The Future of the Past," pp. 16-19.

times. But it will not make us rooted. Rootedness is a state of unconscious at-homeness, which is probably beyond the grasp of most people in a highly mobile, rapidly changing, modern society.[30] Lynch does not endorse traditional landmark preservation that values only the oldest or most beautiful. He prefers an emphasis on the immediate past--the past experienced by those still living. Decisions about what to save should be based not so much on the opinions of specialist advisers as on "the memories and hopes of the users of an area."[31] The landmarks and the landscapes of people's own past lives should form the basis of an effort to manage the sense of a place. Lowenthal has noted that at many places in America "the valued past is merely museumized, not integrated with the present."[32] Yet he has shown greater sympathy than Relph for such modes of preservation: "better . . . a light-hearted dalliance with the past than a wholesale rejection of it."[33] Such an appreciation for even the more bizarre and contrived forms of historic preservation is based on recognition of the fact that all forms of historic preservation, even those that appear the least meddlesome, remake the past and our sense of it.

Despite the caveats, there is general agreement that the desire to maintain and enhance a sense of place underlies much modern-day preservation activity. However, examination of particular senses of place and how they motivate preservationism are few. One might think that the preservationists themselves would articulate the sense of place they are defending. But such is not often the case. In our study of Sacramento preservationists, we confirmed that concern for a traditional sense of place was among the most important reasons people gave for becoming involved in historic preservation activities.[34] Yet the Sacramentans' sense of place remained only a shadowy source of inspiration--unexplored and unarticulated. Similarly, from a survey completed by 199 preservation groups in the Philadelphia, Washington, D.C., and San Francisco metropolitan areas, Datel found that the desire to retain distinctive environments and psychological links with the past were important aims of the organizations. But in the literature they produced--such as walking tour pamphlets, local histories, and

---

30. Tuan, "Rootedness versus Sense of Place."

31. Lynch, *What Time Is This Place?*; Lynch, *Managing the Sense of a Region*, pp. 188-92.

32. Lowenthal, "The Past in the American Landscape," p. 110.

33. David Lowenthal, "Conclusion: Dilemmas of Preservation," in *Our Past Before Us,* ed. Lowenthal and Binney, p. 232.

34. Datel and Dingemans, "Historic Preservation and Urban Change," p. 245.

newsletters--expressions of sense of place and discussions of the meaning of places to members and citizens were meagre. They were most likely to appear vis-à-vis great public monuments of obvious symbolic value (what Tuan has called "public symbols") such as Independence Hall or Mount Vernon, rather than workaday landscapes (Tuan's "fields of care").[35] All this is not terribly surprising. Most preservation battles have been rear-guard actions, focused on buildings in imminent danger of being destroyed or mutilated. Typically, there was not much time for preservationists to ruminate on why they felt so strongly that a particular structure or tree or vista ought to be saved. But during the last decade, historic preservation has attracted increased attention and resources. These have made possible many comprehensive surveys of cultural resources and opportunities for reflection apart from the heat of conflicts over particular buildings or neighborhoods.

When one examines the products of such comprehensive survey efforts, however, one is apt to be disappointed by the absence of discussion of local or regional senses of place. Typically such documents are catalogs of structures deemed to be of architectural interest or historical significance. Experts in history and architecture have rated the inventoried buildings based on such criteria as age, association with important events or persons, visual prominence, and architectural excellence. Sites associated with famous events and people and distinctive architecture are important to the senses of place of many people, including, of course, preservationists, who are apt to be well-educated in those matters.

But the objective judgments of an outsider are not the same as the attachments of an insider.[36] Insiders may use the technical language of the architectural historian to make their case to the planning commission or the preservation board, but it is not just the rare eighteenth-century facade or the innovative subdivision plan that ties them to a building or neighborhood.[37] What is the

---

[35]. These terms are used by Tuan in "Space and Place: Humanistic Perspective."

[36]. Notions of insideness and outsideness, whether those labels are used or not, are central to most discussions of sense of place. Insiders, who are intimately acquainted with a place, will have a very different sense of it than outsiders. Insideness and outsideness are discussed in Relph, *Place and Placelessness*; Anne Buttimer, "Home, Reach, and the Sense of Place," in *The Human Experience of Space and Place*, ed. Anne Buttimer and David Seamon (London: Croom Helm, 1980), and David Seamon, "Newcomers, Existential Outsiders and Insiders: Their Portrayal in Two Books by Doris Lessing," in *Humanistic Geography and Literature: Essays on the Experience of Place*, ed. Douglas C. D. Pocock (London: Croom Helm, 1981), pp. 85-100. David E. Sopher, "The Landscape of Home: Myth, Experience, Social Meaning," in *Interpretation of Ordinary Landscapes*, pp. 129-149 discusses related ideas.

[37]. The idea that people with a deeply felt sense of place adopt a technical language when discussing it because no adequate experiential one exists

broader and deeper sense of place that inspires the myriad preservation activities of today? What are the roles of the physical (relating to the way things look) and the socio-psychological (relating to familiarity established through experience) dimensions of sense of place?[38] Are some preservationists defending a traditional sense of place so that it suddenly becomes "historic"? How do particular historic preservation actions (e.g., distributing a newsletter, conducting walking tours, restoring Main Street facades) change the sense of place of public symbols? Of fields of care? How much preservation activity reflects love of the old, and how much fear of the new? To answer these questions would require using several of the techniques explored by geographers and other students of environmental perception. Examination of regional or local literature (fiction and non-fiction) and art;[39] participation in and observation of relevant decision-making groups;[40] questionnaire surveys;[41] interviews;[42] and perhaps cognitive mapping can offer valuable insights.[43] A combination of these admittedly diverse methods would provide the richest and most reliable portrait of the sense of a place. Not only could these approaches help scholars learn more about people's sense of place, but they would complement the cultural resource assessments of experts and provide additional

---

is explored in Linda H. Graber, "Development Control and the Sense of Place: Experiential Foundations of Contemporary Land-Use Planning Movements" (Ph.D. dissertation, University of Minnesota, 1979).

38. The physical and socio-psychological dimensions of sense of place are described in Douglas Pocock and Ray Hudson, *Images of the Urban Environment* (London: Macmillan Press, 1978), pp. 80-86.

39. See Yi-Fu Tuan, "Literature and Geography: Implications for Geographical Research," in *Humanistic Geography,* ed. Ley and Samuels, pp. 194-206, and Pocock, ed., *Humanistic Geography and Literature.*

40. This method is advocated by Edward Gibson in "Understanding the Subjective Meaning of Places," in *Humanistic Geography,* ed. Ley and Samuels, pp. 138-154.

41. Many different kinds of survey questionnaires have been used to gather information on perceptions and attitudes toward places. Some have produced tidbits concerning the historical environment and could be used to discover more. See, for example, Jacquelin A. Burgess, "Stereotypes and Urban Images," *Area,* 6 (1974): 161-71; Theodore W. Jackovics and Thomas F. Saarinen, "The Sense of Place: Student Impressions of Tucson and Phoenix," *Arizona Review* 27 (April 1978): 1-12; David Lowenthal and Marquita Riel, *Publications in Environmental Perception* (New York: American Geographical Society, 1972); and (using the same methodology as the latter) J. A. Burgess and G. E. Hollis, "Personal London," *Geographical Magazine* 50 (1977): 155-159. Other examples can be found in *Images of the Urban Environment,* ed. Pocock and Hudson.

42. Graham D. Rowles, "Reflections on Experiential Fieldwork," in *Humanistic Geography,* ed. Ley and Samuels, pp. 173-93; Kevin Lynch and Malcolm Rivkin, "A Walk Around the Block," *Landscape* 8 (1959): 24-34.

43. For a recent overview, see Roger M. Downs, "Cognitive Mapping: A Thematic Analysis," in *Behavioral Problems in Geography Revisited,* ed. Kevin R. Cox and Reginald C. Golledge (New York and London: Methuen, 1982), pp. 95-122. For a lively example of this enormous genre see Stanley Milgram and Denise Jodelet, "Psychological Maps of Paris," in *Environmental Psychology,* 2nd edition, ed. Harold M. Proshansky, William H. Ittleson, and Leanne G. Rivlin (New York: Holt, Rinehart and Winston, 1976): 104-24.

guidance to people concerned with taking care of their surroundings.

Such approaches are absent because the kinds of experts usually hired to assess cultural resources are not familiar with them and because of the expense. But perhaps they are avoided also because of political conflicts that might be exposed. As Cybriwsky's and Levy's studies of gentrifying neighborhoods in Philadelphia have shown, the architectural features and historical associations that appear to dominate the upper-middle class newcomers' sense of place do not have the same significance to working-class oldtimers whose established social ties and years of experience in the neighborhood are key components of their sense of place.[44] Holdsworth has discussed the competing senses of place of middle-class native-born and working-class immigrant households in a Toronto neighborhood of Victorian houses.[45] Here both groups are newcomers, each trying to establish a sense of place in its own way. The natives prefer restrained exterior restoration, while the immigrants are more apt to make bold changes and use very bright colors and religious icons. Neither look is historically "correct" (although the natives are calling for historic district controls), yet even if one were, should that one be given preference by city regulations? In cases such as these it is easier for politicians to follow "objective" expert advice, rather than try to cope with subtle and conflicting relationships to the environment. But historic preservationists are already becoming more involved in places that are fields of care rather than public symbols; conflicts are unavoidable. Environmental perception scholars have an opportunity to both study and influence that involvement and so steadily remind preservationists, city planners, and developers of the complex experiential nature of sense of place.

Comparative Studies of Historic Preservation

In the above section we called for an in-depth study of sense of place and historic preservation in a particular area. Here we suggest that our understanding of these subjects also can be enhanced by means of comparative, interregional or international studies. A start has been made on this type of work. Travis compared the preservation orientations of regions of the United

---

44. Roman Cybriwsky, "Social Aspects of Neighborhood Change," *Annals of Association of American Geographers* 68 (March 1978): 17-34; Paul R. Levy and Roman A. Cybriwsky, "The Hidden Dimensions of Culture and Class: Philadelphia," in *Back to the City*, ed. Shirley Bradway Laska and Daphne Spain (New York: Pergamon Press, 1980): 138-55.

45. Deryck Holdsworth, "Heritage Fabrication and High-tech Chic," paper presented at the Annual Meeting of the Association of American Geographers, San Antonio, Tex., April 26, 1982. The abstract of this paper is in *AAG Program Abstracts* 1982, pp. 10-11.

States, using listings from the National Register of Historic Places.[46] Newcomb compared preservation planning in the U.S., the U.K., and Denmark, with special reference to recreational sites.[47] His emphasis, however, was on the practical aspects of preservation laws and techniques, rather than on attitudes and perceptions. Holtzner contrasted the preservation of historic urban forms and ways of life in German cities to the American willingness to modify cities for convenience and efficiency.[48] He attributed this difference to Europeans' greater sense of insecurity and pessimism about the future, which were rooted in the disastrous modern history of the region. In the dozen years since Holtzner wrote, Americans have been faced with unprecedented defeats and traumas, and we too have turned toward the solace of the past.

The work of Lowenthal and Prince contrasts perceptions of the past and modes of preserving in Britain and America.[49] According to Lowenthal's analysis, the American way of history has been to isolate, purify, and ossify it in order to use it to convey strictly controlled social messages about American values and virtues. Americans have not hesitated to use history for money-making purposes. Britons, on the other hand, have preferred their historical places to look lived-in and have better integrated their landmarks into the working landscape. They have been more reluctant to exploit commercially their historical places, taking the attitude that hawking history somehow debases it. Perhaps this dichotomy has lessened in some respects, as American preservationists have become increasingly concerned with ordinary towns and neighborhoods that have little to do with tourism.

Our own work on historic preservation utilizes comparative approaches. Dingemans studied interregional variations in the extent and nature of preservation activities in rural northern California.[50] Differences appear to be related to greater appreciation for the hilly landscapes of the Mother Lode, where the Gold Rush is still vividly recalled in the mind and on the ground,

---

46. Richard W. Travis, *Regional Components of the Recognition of Historic Places*, Occasional Publications Paper no. 3 (Urbana, Ill.: University of Illinois, Department of Geography, 1972).

47. Newcomb, *Planning the Past*.

48. Lutz Holtzner, "The Role of History and Tradition in the Urban Geography of West Germany," *Annals of the Association of American Geographers* 60 (1970): 315-39.

49. Lowenthal, "American Way of History" and Lowenthal and Prince, "English Landscape Tastes."

50. This research was supported by a grant from the Kellogg Foundation and the University of California, Davis Public Service Research and Dissemination Program.

than for the flat treeless expanse of irrigated large-scale farming in the Central Valley. Another area where historic preservation is well-established is in the San Francisco Bay Area northern exurbs. Like the Mother Lode, the landscape is hilly, and farms, such as those of the grape-growing areas, are small; the feeling of intimacy so characteristic of many English rural scenes is there. These areas are stimulated to action by urban pressures and can draw upon the expertise of well-informed and politically experienced exurbanites. In contrast, the counties in the Central Valley with low levels of preservation activity lack the stimulus of threatened change and large numbers of sophisticated preservation advocates. More detailed case studies will address the issue of how much of the observed differences in preservation activity is due to preferences for certain kinds of cultural landscapes over others and how much is due to such factors as accessibility, degree of threatened change, presence or absence of leadership, and the like.

A current research effort by Datel comparing historic districts in five metropolitan areas--Philadelphia, Washington, D.C., San Francisco, London, and Paris--notes the growing similarities between preservation tastes in those regions. Designation is applied to the same general kind of places: the civic core highly charged with public meanings; the earliest surviving residential quarters, particularly those with aristocratic associations; ancient villages (whether still physically distinct or having been swallowed by suburban development); garden suburbs and other unified suburban developments; early industrial sites, especially canals and water-powered mills; institutional complexes of attractive or innovative design or historical importance; houses and grounds of the upper class; and gardens, parks, farms, cemeteries, and other open spaces. Differences between metropolises in the categorical distribution of historic districts are in great measure explained by differences in resource base, although administrative, political, and economic factors also impinge. For example, among London's 404 conservation areas, about 175 could be classified as suburban residential developments, mostly from the late nineteenth and early twentieth centuries. Few such areas around Paris have been similarly recognized. To a great extent this difference is explained by the smaller extent and lesser quality of such development around Paris. Similarly, Pennsylvania's village settlement pattern means many village historic districts for Philadelphia, while the plantation system of Virginia means that an especially large share of Washington's historic districts are of the "house-with-grounds" variety. However, despite the variations in built inheritance, the same qualities of human scale, antique texture, harmonious architecture, mature landscaping, romantic associations, and traditional appearance recur again and again. Which of these are universally favored features of valued landscapes?

The five cities do treat differently the places that they designate as historic districts. The cities and countries vary as to the specific preservation powers granted to government authorities and the sticks and carrots with which owners and residents are encouraged to protect and enhance their historical properties. Americans favor the carrot (e.g., tax incentives for certified restorations of historical buildings), since they are less willing than most Europeans to tolerate the stick of restricted property rights. The central cities of Philadelphia and San Francisco have not advanced very far in the local designation and regulation of historic districts, but Washington, D.C., and the European cities with their stronger planning traditions have tough preservation controls. Approaches to physical change in historic districts vary widely, but the variation is more significant within each city than between them. All the metropolitan areas contain examples of historic districts that became decayed and are being restored, others that have long been stable, and still others where new development is permitted, but controlled. Americans have been accused of being peculiarly preoccupied with restoring places to their "original appearance," but the French *secteur sauvegarde* (protected sector) program aimed at restorations that surely were as thorough and extravagant as anything conceived by the Americans. Massive physical changes, while happening at different times and places in each of the cities, have had the uniform effect of uprooting citizens and creating an awareness of history as a different time and of historic landscapes as special places.

These are some of the similarities between preservation in the U.S., Britain, and France. If one reads the manifestos of American, British, and French preservationists one detects close agreements as to the kinds of places that should be preserved (we wonder if the same can be said of the opinions of the "average citizen" of each country--here is another question that merits study). This is a natural result of the tremendous broadening of vision concerning what is worthy of preservation. Further broadening of this vision to encompass the unlovely but beloved could transform historic preservation into a general approach to humanistic place-making.

Environmental perception scholars can play a role in this process by exploring what is still largely *terrae incognitae mentis*--the sense of place of ordinary people.

CHAPTER 9
WHERE DO WE GO FROM HERE ?
A COMMENTARY

Larry R. Ford

The revolution is over. Environmental perception has become an important approach in a wide variety of geographic studies. Indeed, as Saarinen has pointed out, a great many geographers have integrated perception approaches so thoroughly and completely into their research interests that they do not claim to be concerned with environmental perception per se. This is as it should be. Environmental perception should not be thought of as a discipline or subdiscipline but rather as valuable approach toward gleaning insights within a number of existing disciplines. As several of the authors demonstrated, perception is now thriving not only in the traditional areas of psychology, design, and geography, but also anthropology, sociology, and political science. In addition, a number of theoretical frameworks for further perception studies have been identified which should serve to organize inter-specialty and inter-disciplinary cooperation. In short, things seem to be going pretty well. Nevertheless, there are some concerns that we as geographers interested in perception should be aware of lest we suffer as well as gain in the new enthusiasm for the approach.

First, we must guard against becoming too esoteric, too inter-disciplinary, too isolated from the traditional concerns of geography. If we become overly concerned with psychobiological or cognitive dimensions of perception, for example, we could eventually find ourselves in a situation analogous to that of the extreme "quantifiers" of the 1950's and 1960's. "Geography as geometry" and "geography as recreational mathematics" became too remote from the concerns of most geographers, and perhaps actually retarded the acceptance of reasonable and useful quantitative techniques. We must endeavor to translate breakthroughs in perception research so as to make them relevant and useful to a large number of geographers. The environmental perception specialty group should play an important role in this filtering process, i.e., keeping tabs on the latest conceptual and methodological developments and modeling them to suit geographic purposes.

A second concern is that geographic research in environmental perception will be overwhelmed and engulfed as enthusiasm for the approach spreads to an increasing number of larger disciplines. Geography is a small discipline. If we attempt to keep up with and do research in a wide variety of perception specialties, we may well spread ourselves too thin. We cannot be all things to all people and our credibility may be stretched to the limit if we invade territories that we have no established background for dealing with. Interdisciplinary cooperation implies that each discipline has something to offer the other and so geographers should make sure the core is adequately developed before engaging in too many forays into peripheral research themes. Such forays are often fun and should come later, but first we must establish that we (as geographers) have something valuable to contribute. What then constitutes the core of perception geography? I offer a few suggestions.

Geography has four traditions that are exceptionally appropriate for the environmental perception approach. These are (1) Landscape, (2) Spatial, (3) Man-Land, and (4) Comparative. The Landscape tradition dove-tails to some degree with what Sell, Taylor and Zube refer to as the Experiential. Geographers have a long and well-developed research tradition of "reading" the landscape so as to develop process-oriented hypotheses. The works of such people as Sauer, Lowenthal, Meinig, and Tuan epitomize this tradition. More recently, work by geographers on historic preservation, summarized by Datel and Dingemans, has served to focus the environmental approach upon a particular landscape phenomenon, i.e., what does the sudden concern for recycling the built environment tell us about the processes shaping our modern urban fabric--our values, economic system, energy shortages, demographic trends, etc. When buildings and entire neighborhoods once perceived as "obsolete," "slummy," and "dangerous" suddenly become "human scale," "delightful," and "interesting," there is obviously something going on that is of interest to geographers concerned with the ways in which people relate to place. Historic preservation and the associated issues of gentrification, urban revitalization, ghettoization, and urban models thus provides a current focus for geographic environmental perception studies dealing with "reading the landscape" and "landscape tastes." These are themes that geographers know something about.

A second tradition in geography that can provide a focus for environmental perception studies is the spatial. Geographers, for the most part, have abandoned the optimizing economic man in favor of a more complex human being making a wide variety of satisficing decisions based upon his/her perception of reality. Geographers have a long tradition of dealing with location decisions and spatial behavior--indeed, many would argue that this is the real core of the discipline. Zannaras summarized many of these ideas in discussing the relationship between place perception and spatial behavior in San Antonio. All behavior may be a result of perception rather than objective reality but this seems to be especially true in the realm of spatial behavior in which information about "the other side of town" and/or "good areas and bad areas" tends to be highly subject to the whims and vagaries of random encounters with the landscape. Spatial behavior is and should be an area through which geographers can make a real and lasting contribution to environmental perception research.

The original boom in interest in perception in geography was based upon one of its oldest traditions, Man-Land. Studies of the human response to perceived natural hazards provided an ideal focus for geographers to combine and integrate knowledge of physical processes and knowledge of spatial behavior through the medium of environmental perception. This duality in geographic wisdom allows geographers to contribute to an understanding of reality as well as the human response to that reality, i.e., provide a context for the perception process. There is a tendency for geographers to broaden their interests and to examine responses to a wider variety of hazards (social, political, etc.) but perhaps we should stick to what we do best. We have reached the critical mass in this research area to make a solid contribution to others via interdisciplinary cooperation.

A final core tradition in geography is the comparative study or "areal differentiation." People and places differ and models developed in other disciplines only rarely take this axiom into account. People may well perceive their environments differently depending upon their cultural backgrounds, social backgrounds, personal histories, and the character of the places they have experienced. Once more geographers are in a position to contribute. Even if we assume that it is difficult to compare the perceptions of very different cultures (an assumption I do not make), there is still ample opportunity to compare different groups within a culture and similar groups located in different types of places. Hart, for example, demonstrated that geographers can play an important role in helping developmental psychologists explore the perceptions of children. If we are to specialize and make a major contribution in this area, however, perhaps the place to start would be in the comparison of children growing up in the Bronx or Manhattan with those growing up in the desert. How do they perceive reality? How do they process information? How do they learn to behave spatially? How do they learn to read the landscape? Geographers, as a group, know something about regional cultural traditions and this is a chance to use that knowledge.

I agree with Saarinen that there is reason for optimism as the environmental perception approach has become well-integrated into traditional geographic pursuits. There is reason for even more optimism if we can utilize the insights of an increasingly interdisciplinary enthusiasm for environmental perception in the development and enhancement of a few core areas in which geographers can make a solid contribution to the accumulation of wisdom with regard to the perceived world.

CHAPTER 10
MODESTY AND THE MOVEMENT
A COMMENTARY

James M. Blaut

Our chairman has lengthened my leash a bit so that I may comment on the field of environmental perception (or whatever else we may choose to call it) as a whole.[1] A field, yes, but also a movement, what Ludwik Fleck would have called a "thought community," with its own "thought style".[2] The environmental perception movement seems to me to be a product of the epoch and *Zeitgeist* called "the Sixties." Environmental perception as a research field, on the other hand, is mainly a product of special conditions which prevailed in our science and some others a few years earlier: in the middle and late Fifties, to be precise. This was a time when old viewpoints were being discarded in geography because they didn't seem to be able to solve the important problems of theory and practice. Many of us felt that, if somehow we could get inside the minds of the human beings whose environmental actions we were studying, if we could uncover their ideas, preferences, motivations, we might thereby find out the reasons--the scientific explanations--for the larger facts and processes that were our concern as geographers. So we turned for help to psychology and anthropology. But we found, to our surprise, that these other scientific fields, and some client fields like architecture and planning, were engaged in much the same quest and were no closer to solutions than we were. It quickly became clear that human environmental behavior, its internal (mainly psychological) attributes and external (mainly artifactual and spatial) linkages, constitutes a legitimate, important, and still largely unexplored field. Research and practice in this field has steadily progressed over the past twenty years or so, and nobody needs to feel ashamed of what has been accomplished.

The movement is something larger. It is rooted in the axiomatic belief that the study of individual, concrete, human beings and their mental (and spiritual) attributes, or their

---

[1]. This essay is an expansion of a presentation given at the "Environmental Perception: Inventory and Prospects" session of the 1982 Annual Meeting of the Association of American Geographers, San Antonio, Texas, April 28, 1982.

[2]. Ludwik, Fleck, *Genesis and Development of a Scientific Fact* (Chicago: University of Chicago Press, 1979, originally in German, 1935).

biological make-up, is somehow one, the true pathway to explanation in geography. This is a radical reductionism. Processes at the scale of social collectivities and systems, communities, cultures, nations, and so on, can only be explained, it is argued, if we reduce them to processes at the scale of the individual human being, to psychological or biological events which are considered by some to be empirically discoverable and by others to be beyond science--to be metaphysical--hence matters of the spirit or perhaps the soul. This radical reductionism is of course not at all new in science and philosophy, and it has been put forward from time to time as an organizing schema in geography. But what transformed this psychologistic and sometimes biologistic reductionism into a social movement was the Sixties. That was a time when science as a whole was under general attack for failing to solve human problems--worse: for failing even to notice human problems. A number of radical epistemologies and methodologies emerged from this critique. One of these, broadly described, was a demand that science return to the real, feeling, apperceiving, human being, because here was the locus of suffering--in Vietnam, in Mississippi, in Haight-Ashbury and the college dorms, in ghettos and ill-designed public housing projects, but always in real, discrete, human bodies and minds.

I think that this social, even political, process explains a lot of what has happened since the Sixties in the study of environmental perception or--the term I prefer--psychogeography. One very positive effect of the movement was the tremendous growth of interest in this new field, hence a certain amount of recruitment and a great deal of solid accomplishment. But the positive consequences cannot have much outweighed the negative, because the axiom of radical reductionism was itself mistaken, and the scientific and practical consequences of work based in this axiom had to be inadequate. Many of us found all of this out in a series of what can be thought of as stages of disillusionment. The more we studied human individuals, adults and children in our own society the more we realized they are not explicable *sui generis*; and are not free--that is: are not able to realize the environmental actions which their perceptions, knowledge, and values would lead them to carry out. Therefore our predictions and explanations were failing, and our prescriptions for change in the external environment were not improving the human condition.

What we have, I think, are two levels of aspiration. The more modest level is characteristic of psychogeography (or environmental perception) as a research field, not a movement. Valuable work in this field can be carried out by investigators who ground their ideas in any of a number of different epistemologies and social or political axioms. Two basic kinds of research in particular are important and are partially--but only partially--indifferent to questions of epistemology and world view. One is the straightforward study of individual environmental behavior and its psychological attributes. I mean that we can probably do our work on this subject matter with neither more nor less philosophical and sociopolitical contamination than one finds in the science of psychology, one in which the material environment is more salient than the social environment. Solid research is carried on by psychologists holding many different sets of axiomatic principles; the same--neither more nor less--should be true of ourselves.

The second kind of research at this modest level of aspiration can best be described by analogy. Imagine a burglar who wants to gain entrance to a tall apartment building so that he can burglarize one of the upstairs apartments. Imagine that he climbs the fire escape, testing windows and doors at each floor, and eventually finds that he can only enter the building through the basement. In other words, his goal is upstairs but to reach that goal he must enter downstairs. The analogy is very direct to someone who wishes to study human processes at a high level of aggregation or system--the level of communities, nations, states, cultures, etc.--but finds it necessary, as a matter of research strategy or method, to examine these processes at the level of the single individual or small group. No judgment need be made here as to whether social processes throughout are reducible or not to individual processes; it is merely true that some of the large-scale processes, sometimes, can be examined more feasibly (or only) at the level of the individual and his or her geographical milieu. Thus psychogeography represents a microgeographic methodology which can provide a great deal of knowledge about macrogeography, knowledge which cannot, as Robert Platt pointed out a long time ago, be obtained through macrogeographic research. Perhaps the best example of this strategy--Hart makes this point very effectively in his important paper at this gathering--is the study of children's environmental perception and cognition, learning and values as a means not only of understanding the psychogeography of children but as a means of understanding these matters at all levels of a culture.[3] For, however we choose to define culture, holistically or

---

3. See chapter 7.

psychologically, all of it must be transmitted from one generation to the next, and many parts of culture which are otherwise difficult or impossible to study are peculiarly accessible during the transmission process. It is, so to speak, the open window which gives us access to all floors of the building.

The immodest level of aspiration is one which claims that basic problems, scientific and social, can only be solved if we reduce them to the level of the individual human being and his or her biology or psychology or spiritual existence. Conversely, it claims for our field an ability to solve such basic problems, problems ordinarily attacked at a higher level of aggregation or system--sociological, economic, political, and so on--by reducing them to the level of the individual. It claims to be able to slough off the layers of culture and society and reach down to the real "human nature." And making this immodest claim, it becomes intensely positivistic in the sense of asserting that it possesses research designs which do indeed probe down to the level of "human nature," designs which ignore or discard the cautions and safeguards against elitism and ethnocentrism.

The two basic reductionist approaches are biologism and psychologism. Biologism has the general form of asserting that human behavior operates on some clearly distinguishable infra-cultural or pre-cultural biological basis. There are several candidates. One is, of course, genetics, seen mainly as a supposed determinant of intelligence or rationality, and the claim that genes determine intellect and behavior is made so regularly in the context of comparison among human groups, some asserted to have good genes and some bad, that genetic biologism can properly be bracketed with classical racism and discarded. Enough said about it. But other forms of biologism are more subtle and hence more inclined to be admitted into our work. Everyone knows of the love affair which the perception movement had, and I suppose still has, with the field called "ethology" and with writers like Robert Ardrey, J. B. Calhoun, Konrad Lorenz, and E. O. Wilson (the "socio-biologist") who claim to have found infra-cultural determinants of human psychology, of human behavior. Some of the candidate-determinants are: the uncontrollable sexual urge deriving from our animal ancestry (and animal nature) which leads us to overcrowd and overpopulate the earth; the instinct of aggressiveness which compels us to behave in a competitive, bellicose way toward one another; the supposed biological need for "territory" which is claimed to be the basal explanation for wars (with help from the instinct of aggressiveness); and to be the biological, evolutionary basis for the urge to possess private property, which thus becomes a product of human nature, not of history; and to be the biological explanation (via "crowding") for the problems of ghettos and prisons, as well as the argument that all such problems are so deeply rooted in our biology as to be incapable of solution by social and political means.

These putative biological determinants of behavior fail for rather simple reasons. Liberals and socialists may simply note that all of the supposedly "natural" forms of behavior, those said to have bases in biological evolution and its ontogenetic outcome, are behaviors of the sort either positively valued by present-day capitalist society or declared by the ideologists of this society to be unmitigable, because "natural." But this is of course merely a post hoc argument. More telling is the fact that essentially all of the evidence adduced to support these arguments about infra-cultural determinants of behavior has been drawn from a few selected species of animals which display, and hence illustrate, these behaviors. But one can choose other species to supply evidence, if one wishes it, that very different behavior is at least equally "natural." Thus, ethology and socio-biology seem to be merely a sophisticated form of totemism, selecting the appropriate totems for the chosen behaviors. Before leaving this matter I must note that ethology and socio-biology are not, as their advocates claim, new fields of knowledge. They are fractional throwoffs from other fields, principally animal ecology. Mainstream animal ecologists like A. E. Emerson, W. C. Allee, and T. C. Schneirla have tended to follow a different theoretical tendency, one which sees cooperative behavior and truly social (or, if you prefer, "altruistic") patterns of interaction as having at least as crucial a role in animals as competitive and aggressive ones. To deny this mainstream theory, it seems, you must invent a new science and give it a new name. I wish that psychogeographers would pay as much attention to what Allee called "general sociology" as they do to "socio-biology," and as much to what Schneirla called "comparative psychology" as they do to "ethology." There indeed are biological bases to behavior, but they are probably not tooth-and-fang instincts, they are inextricably enmeshed in cultural behavior, and--perhaps most important--they are as yet hardly known. Biologism seems to give our work a marvelously scientific aura, but it is in reality quite unscientific.

Psychologism, or psychological reductionism, is the second immodest approach to work in our field, immodest because it claims for us an ability to explain, predict, and cure beyond the limits to which our own field, and our own competence, can take us. I would distinguish three forms of this immodest approach, of which the first is perhaps less a matter of psychology than of philosophy. It is an approach which I view with respect, because it is usually put forward as a coherent body of ideas, carefully and explicitly grounded in one of several a priori philosophical bases which are then connected logically to propositions about method and theory. But if you do not accept the philosophy, as I don't, you tend not to accept the scientific propositions. A number of philosophical systems (or at least viewpoints) are involved, and I will not try to characterize them individually. All postulate that the category "mind" is either radically disconnected from the external world or is respondent in some senses and in some way to a beyond-physical, "meta-physical," cause, a spirit or perhaps soul. Hence, the ultimate cause of behavior must lie in (or beneath) mind itself, not in external circumstances or group characteristics or culture. And mind itself is not seen (as cultural anthropologists and cultural geographers generally see it) as a product of the impact of learning on the biological individual. Learning, the resultant of external impinging influences, is thought merely to be a part of the explanation and perhaps not an important part.

The second form of psychologism can be called "cognitive positivism." It is very popular in our field. Unlike the form discussed above, it tends rather ostentatiously to deny the relevance of metaphysics. This viewpoint postulates a rational faculty which processes, cognitively, perceptual "information" to arrive at "decisions," which are then, in their turn, presumed quite arbitrarily to be effectuated in behavior and thus to have effective consequences in the external world. Central to this viewpoint is the idea of rationality, often described as "bounded rationality," or as rationality struggling to distance itself from the emotions, passions, or urges. The concept has a direct ancestry in Weberian (and other Neo-Kantian) notions of rationality, and in classical economics, as well as in more recent notions about decision-making. The idea of rationality then leads to the postulation of an autonomous intelligence, influenced by external and internal (emotional) forces but not caused by them, and hence ultimately an individual possession, not a product of society and environment, and thus a faculty which works toward individual, not social, goals. Typically (but not always) it is claimed also that this rational ability, this cognitive intelligence, is not found in equal measure among all human groups and taxa--cultures, classes, age groups, and perhaps also genders. This, as we will see, is a very troublesome issue.

The third kind of psychologism can be called "subjective perceptionism." Not only is it popular, but it seems to be pretty much what the masses of geographers at large, or a goodly share of them, think of as "the perception movement," or "perceptionism," or even "behavioral geography" (They are wrong). This viewpoint is in a way the reciprocal of the former viewpoint, cognitive positivism, and, like it, is usually (but not always) detached from metaphysical assumptions and indeed from explicit philosophy in general. Subjective perceptionism is grounded in this postulate: People's behavior in the environment is *not* rational. Or, to be more precise, the reason for studying people's perception of the environment is grounded in the fact that people do not behave in the environment as would be expected if they simply followed rational decision-making processes, processes generally equated with the decisions recommended (etically) by experts.

Now a careful distinction needs to be made between the reductionist, psychologistic, viewpoint which I describe as subjective perceptionism and the perfectly valid viewpoint which considers all of the perceptual, cognitive, and evaluative processes which underlie (that is, precede and flow into) overt behavior, which does not commit the error of imagining that people possess perfect knowledge of an environmental situation and the companion error of imagining that people act more-or-less automatically on the signals received from the environment. Subjective perceptionism expects irrationality. Its project calls for an analysis of the cultural bases of environmental behavior, that is, the kinds of values and other attitudes, plus the selective perceptions, which are internalized from a person's culture through learning. But its project also, and more fundamentally, calls for a probing into the inexplicable, idiosyncratic, truly subjective bases for behavior, bases not in principle derived from a person's culture, hence purely psychologistic: a cause uncaused. Even this expectation of the capricious and irrational would not deserve (much) criticism were it not for the distorted research methodologies which this viewpoint has sired. Not expecting rationality, it does not probe deeply enough to see whether rationality is indeed present. Not expecting a logical basis for behavior in culture, it does not examine the culture with sufficient care. Thus, it finds exactly what it expects to find: pure subjectivity; pure caprice. But again the problem is to be found in the consequences. In research practice, and particularly in other cultural research practice, subjective perceptionists employ a model of the environmental actor as somehow governed by "traditional" attitudes, inappropriate to the existing environment and the problems it poses, attitudes which are often thought to reflect the atavistic survival of a "traditional culture." It happens that there exists an extremely crude theory about "traditional societies" with which is associated a "traditional mind" or "primitive mind," a theory which can be, and regularly is, plugged directly into subjective perceptionism to give us a pre-packaged theory to the effect that those whom we study act inappropriately in the environment because of their "traditionalism." And all of this is then deployed in methodologically crude but ambitious studies which try to compare different groups of people, often at opposite ends of the earth.

Thus, subjective perceptionism, particularly (but not only) in its cross-group comparisons, becomes a means by which ethnocentric and elitist mythology presents itself as science. What we end up with in much of this literature is really a modern version of the classical Weberian view that there is, on the one hand, the rational mind of the Northwest European Protestant burgher and, on the other, the subjective, traditional mind of everyone else; and that this fact of selective rationality explains all of modern history.

Two matters remain to be dealt with. One is a cataloguing of the most serious problems which can be traced to radical reductionism (attending mainly to its psychologistic forms). The second is a very brief consideration of the relationship between my critique of reductionism and the eternal philosophical debates between "reductionists" (also called "methodological individualists") and "holists." Probably the most serious errors which result from reductionism in its psychologistic and biologistic forms are those which ascribe causal efficacy to human individuals and not to human collectivities; more completely: to isolated individuals and not to social systems like communities, cultures, and states. This leads to a blindness toward all causes of environmental behavior which are external to the individual. We tend to postulate or assume that there are important sources of the individual's environmental behavior which do not emerge from learning and hence from culture and the external environment. So we may impute unwarranted causal efficacy to supposedly innate forces, like genes, or like the somewhat mysterious force of "development" (as postulated, e.g., by Piaget, in contradistinction to learning), or like Weberian "rationality," or like caprice.

Now this theoretical individualism is a problem in itself, but it becomes vastly more so because of our methodologies. Innate, individual processes tend always to be detailed as a resultant, a remainder, after we have located the processes linked to culture, to class, to gender, to age, to idiosyncratic environmental experience, and so on. This implies that every (I think) limitation of our methodology leads us to impute more and more importance to the residual category, the individual and presumably innate variables. For example, if I ask someone the right question, I will probably discover a connection between a given behavior and culture or experience. If I do not think of the right question to ask, I will be tempted to think that there is no connection to culture or experience, none having been elicited. And I may then assume an innate, individual cause, or (worse still) I may assume that the behavior is capricious and irrational. This methodological problem reaches its apogee when we try to find out whether an act reflects rational decision-making; if we do not ask the right questions, hence do not learn about the ratiocinative (and learning) processes which led up to an act, we regularly judge the act to be "irrational"--or "traditional" when the judgment is made in what we take to be a "traditional society."

Undoubtedly the most serious consequence of the error just described, the tendency to find causes of environmental behavior in the isolated individual, is plainly political. Unaware of external factors, we assume that individuals are autonomous actors, and can effectuate their decisions regardless of context. This is called "voluntarism." It leads us to believe that there is freedom of action where none may exist. It also leads us to imagine that solutions to environmental problems which are available in affluent communities, like American suburbs, are available everywhere else. And it makes us particularly vulnerable to the ideology which maintains that poverty is the fault of the poor, and underdevelopment is the fault of the underdeveloped, and socio-economic development is really accessible to all societies without the need for major social change because failure to develop is not due to external constraints, political or economic, but to some individual failing, some psychological quirk. In other cultural studies of environmental behavior, this error leads us to imagine that people have available to them a large number of feasible behaviors, what some of us call "available adjustments," when in fact they may have few or none (save emigration).

Another unfortunate effect, or correlative, of reductionism is a tendency to underestimate the importance of culture. I mean here the culture which the individual has internalized through learning and that part of the external culture which operates to constrain or shape his or her behavior as a member of the culture (No invocation here of holism). One attribute of culture is, of course, historicity. Therefore an appeal to infra-cultural forces in the psyche or genes is an appeal to something historical as well as non-cultural: to human nature. Or to animal nature. Or to the "rational mind" (of the European). Or to the "primitive mind" (of someone else). More than incidentally, this lack of attention to period and place and culture is also a lack of attention to class.

Among many other problems deriving from reductionism, only one calls for discussion here. This one should perhaps be described not as a problem but as a peculiar intensification of other problems which occurs in one sort of research context: in studies of environmental perception or psychogeography which are carried out in Third World cultures, studies called "cross-cultural" when comparisons are to be ventured. When we psychogeographers carry out research in our own culture, we take suitable methodological precautions against error, but we are so fully attuned to the resonances of our own culture that some of the precautions are too obvious to require conscious attention and formal recognition (For example, the language of interviews.) When we work in other, very unfamiliar, cultures we have the tendency to take no greater precautions than we do at home, and the results are very bad indeed. I am sure this habit is in part a reflection of the youthfulness of our field of study. But it is also due to the problems discussed here--problems of methodological immodesty. Imagining that we are able to come to grips with real human nature in our work, to probe down beneath the layers of culture and external environment, we do not bother to pay enough attention to culture and environment. And we do not worry very much about the ways in which culture (and also class) must be taken into careful account in our procedures. We do not even pause to inquire whether other fields, notably cultural anthropology (but also cultural geography), have amassed any research experience which would need to be taken into account in our own designs.

The results of all this (and more) tend to be about as follows. The natives we study are judged to be less rational in their use of resources, or their "adjustments" to environmental problems, than we (Westerners) are. The problems themselves are attributed to some combination of natural forces and the failure of the individual to behave rationally; hence oppressive governments, landlords, multinational corporations, and the like are pretty much absolved from responsibility. Irrationality, meaning here failure to make the appropriate decisions vis-à-vis resources, is attributed either to the "primitive mind" theory or to the companion theory of "traditionalism." Note that these judgments are direct consequences of what I have described as the method of residuals: having failed to communicate with a respondent well enough to discover the, usually quite rational, reasons for his or her behavior, we declare that there are no such reasons. We are helped in all of this by our choice of methods which are flawed, or useless, or incapable of delivering any results which might disconfirm our hypotheses--so much for Popper and "falsification"--or dispute our preconceptions.

In discussing this matter I am aware that it needs to be explained in much greater detail than is possible here, and that the critique should be directed at specific work. I will limit myself to two concrete examples, one being a flawed piece of research that I directed, the other being the White-Kates-Burton effort at cross-cultural (or "comparative") research on natural hazards. My work was an attempt to find out why soil erosion was a serious problem in the Blue Mountains of Jamaica.[4] The study involved a two-and-a-half month stay in one village and intensive observation, interview, and environmental analysis carried out by myself and three graduate students (one geographer and two anthropologists), working in our primary language and in a culture and agricultural system with which I was very familiar. All of this needs to be said to emphasize the fact that our data were reasonably comprehensive and reliable. The problem was in the interpretation of these data. Many factors were found to contribute to the erosion problem. Objectively, it was clear that the farmers' knowledge about soil processes and conservation practices was less than complete: they were not omniscient. This convinced me that imperfect perception and cognition was the single most important factor (among many) in inhibiting the deployment of adequate soil conservation practices. I was quite wrong. I should have realized that nobody's perception and cognition of the environment is completely adequate. The proper question to ask would have been: if the farmers knew more than they really do about soil processes and conservation, would there be any less erosion in this area than there actually is? The answer would

---

4. James M. Blaut et al., "A Study of Cultural Determinants of Soil Erosion and Conservation in the Blue Mountains of Jamaica," *Social and Economic Studies* 8 (1959): 402-420.

have been no, and that answer could have been obtained in the course of the research. The key factors were economic and political, not perceptual or cognitive. I was misled by all of the intriguing evidence I was obtaining about the limits of what today we would call the farmers' ethno-pedology, their soil science. Their scientific knowledge was, in general, adequate to put to use any of a number of possible conservation methods, had these methods really been accessible to these farmers. Therefore, in relativistic terms, the farmers were indeed omniscient. The conclusion is: evidence we obtain about perception and cognition is always of scientific significance, but it may explain very little external to its own (modest) sphere of facts. I am now certain that lack of knowledge is very rarely a limiting factor in peasant ecologies; that most peasants' perceptions and cognitions are ample and expansible enough to make use of material resources as these become available--which happens very rarely in capitalist Third World countries. Concerning the error I made in Jamaica, I console myself with the thought that it happened 25 years ago, when I was much younger and so was psychogeography.

The White-Kates-Burton research had the laudable but vain ambition of providing a world-wide cross-cultural comparison of adjustments of natural hazards. My concern is with the incautious, perhaps even slipshod approach taken to several Third World groups studied in Africa and Latin America.[5] The backbone of the research is a questionnaire which is to be administered, with some local adaptation, to mainly rural folk in a number of countries.[6] I am not aware of any prior attempt, anywhere, to design and employ a literally pan-cultural questionnaire more complex than a census schedule, and this one is exceedingly complex, or at least ambitious, with aims to ferret out deep-seated personality traits (like "fatalism," "sense of inner control"), ethnoscience of disasters, probably overt behaviors in future disasters, and other such subtle matters of culture. Saarinen has provided a forceful critique of this questionnaire, but I have some observations to add to his.[7] Anthropologists and anthropologically minded geographers use questionnaires in other-cultural research, but generally as a self-guide for long, informal, unstructured interviews, interviews

---

5. Gilbert F. White, ed., *Natural Hazards: Local, National, Global* (New York: Oxford University Press, 1974), and Ian Burton, Robert W. Kates, and Gilbert F. White, *The Environment as Hazard* (New York: Oxford University Press, 1978).

6. White, op. cit., pp. 6-10.

7. Thomas F. Saarinen, "Problems in the Use of a Standardized Questionnaire for Cross-cultural Research on Perception of Natural Hazards," chapter 23 in White, op. cit.

carried out after long periods spent studying the culture (and language) and establishing rapport with respondents. I see no evidence that the experience of thousands of other-cultural studies, and a large body of grounded theory in psychological anthropology, ethnolinguistics and ethnoscience, and so on, was simply ignored. Anthropologists, also, along with their fellow-travelling geographers, would recommend full training in the relevant culture and language, something attended to for some of the investigators and their field sites but not, apparently, for others. There would be recommendation for long and intensive field experience, again only spotty in this study. There would be further recommendations about the social context in which interviews should take place (with attention to roles, statuses, and of course, language), about the need to fully describe methods and context in published reports, and--very crucially--about the need to take great caution in venturing cross-cultural generalizations, particularly when many profoundly dissimilar societies are to be compared, and compared as to subtle personality traits, hypothetical behaviors not directly observed, and the like.

As was perhaps inevitable, this study produced more than one ethnocentric error. For example, Puerto Ricans are compared with North Americans--apparently the questionnaire was administered in English to both groups--and the former were found to be, in essence, less rational than the latter in their attitudes toward hurricane hazards.[8] They lack "a sense of inner control",[9] that is, an attitude of being able to do something about the situation. They seem to be "less autonomous . . . more fatalistic, more accepting, and yet more frightened of the hurricane threat".[10] In general: this study provided some very good information about some field sites, and it provided invaluable other-cultural experience for a number of geographers, but as a comparative, cross-cultural study, it cannot be accounted a success. I would add that supportable/cross-cultural generalizations in our field of concern, as in most matters involving the human mind, are still largely inaccessible.[11]

---

8. Duane D. Baumann and John H. Sims, "Human Response to the Hurricane," chapter 4 in White, op. cit.

9. Burton, Kates, and White, op. cit., p 107.

10. Baumann and Sims, op. cit., p. 30.

11. This judgment extends to most of the field of so-called cross-cultural psychology. (See in particular the *Journal of Cross-Cultural Psychology*.) A significant part of the work in this field consists of studies by White South African psychologists purporting to show that Black South Africans are inferior to Whites on some psychological measure; and somewhat similar studies by Israeli psychologists comparing Israeli and Palestinian Arab subjects; and so-called

Finally, a word about the philosophers' reductionism-vs.-holism debate, or rather about its essential irrelevance to the issues discussed here. The question we are chiefly concerned with is whether environmental behavior reflects external, learned factors or else some sort of *sui generis* internal cause such as rationality, the spirit, an instinct of aggressiveness, or simple caprice. We know, as a matter of context, that the ideological environment surrounding us (in the United States) will favor recourse to individualistic, voluntaristic models and conclusions. We also know from practice that the influences of culture and experience are hard to uncover, hence our tendency, via the residual bias, toward radical reductionism. I will not repeat what has been said already, but from it we can conclude, I think, that the problem of reductionism is primarily one of methodology and theory, not philosophy. As to philosophy, there is indeed a real and continuing debate about reductionism. However, few participants defend either metaphysical holism of the thoroughgoing (Hegelian) variety or the extreme reductionism which declares that "'army' is just the plural of 'soldier'." Some years ago I argued (in Blaut 1962) that there is as much metaphysical holism in the thesis that the "real" entity is the individual human being as there is in the thesis that the "real" entity is the "culture whole," the state, society, or what-have-you.[12] In both cases we are positing what Whitehead called a "natural entity;" we are imposing something on the world, not ascertaining some thing about the world. An attractive position (advanced, I believe, by Reichenbach) holds that all statements at a given level of aggregation or system can *in principle* be translated into statements at higher and lower levels. This, of course, begs the question whether and when this can be done in practice. All of which brings us around finally to the methodological issue which I have illustrated with the story about the burglar who wants to burgle the penthouse but has to enter the

---

Piagetian replication studies in Third World communities, studies so poorly designed that these people cannot possibly demonstrate cognitive abilities comparable to Westerners (in, e.g., "abstract thought"). Other than this work, there are some very good studies which are other-cultural but not cross-cultural, i.e., not seeking comparative generalizations, and, finally, there are a very few interesting and properly tentative studies which probe the cross-cultural dimension. It is no coincidence that this field tends to be shunned by psychological anthropology and ethnoscience. The moral for ourselves seems to be this: other-cultural studies should be encouraged, but not so comparative studies seeking pan-cultural generalizations at a time when we do not yet know how to design research than can elicit comparable behavior in cultures that are dissimilar to each other and to ourselves. And methodological studies should in general precede the substantive (comparative) ones until we know how to do this kind of thing in a way that is non-ethnocentric, hence advances science, not prejudice.

    12. James M. Blaut, "Object and Relationship," *The Professional Geographer* 14, no. 6 (1962) pp. 1-7.

building at the basement level. In other words: we want to make statements about higher-order systems like communities, cultures, and states, but we have to make do with the raw material at hand: the individual human being and his or her microgeographic environment.

PART III
PHILOSOPHICAL DIRECTIONS

CHAPTER 11
PHILOSOPHICAL DIRECTIONS IN BEHAVIORAL GEOGRAPHY WITH AN EMPHASIS ON THE PHENOMENOLOGICAL CONTRIBUTION

David Seamon

A frequently cited philosophical work in geography today is Derek Gregory's *Ideology, Science and Human Geography,* which argues that current geographical research, physical as well as human, is largely dominated by a positivist outlook.[1] Positivism is the philosophical stance that genuine knowledge must be perceptible in time and space and thus founded on empirical validity.[2] Positivist geography is associated with such terms as quantification, causality, explanation, certitude, prediction, empirical validity, public repeatability and validation. Gregory argues that most of the geographical research arising from a positivist stance is "unexamined discourse" because researchers have not directly considered the underlying assumptions, motivations and world view guiding their work.[3] Gregory calls for a *critical* geography--i.e., a geography that subjects its methods, values and motives of research to continual explicit evaluation. The lack of this critical spirit, says Gregory, makes positivism and thus positivist geography an ideology rather than a science. The first part of his book is a critical evaluation of positivism and positivist spatial science. Gregory concludes that the approach falters badly, leading to a way of knowing which "makes social science an activity performed *on* rather than *in* society, one which portrays society but which is at the same time estranged from it."[4]

---

[1]. Derek Gregory, *Ideology, Science and Human Geography* (London: Hutchinson, 1978).

[2]. Gustav Bergman, *The Metaphysics of Logical Positivism* (Madison: University of Wisconsin Press, 1967); Marwyn Samuels, "Science and Geography: an Existential Appraisal," (unpublished doctoral dissertation, University of Washington, 1971).

[3]. Gregory, op. cit., p. 63.

[4]. Ibid, p. 51.

In the second half of his book, Gregory examines alternatives to the positivist approach. He claims that a revitalization in geography requires a mode of thinking which is three things at once: first, *structural,* in that the geographer relates particular individual and group situations as they are enmeshed in and impacted by larger, less readily apparent cultural and societal processes and strutures; second, *reflexive,* in that the geographer works to make explicit the taken-for-granted qualities of individual and groups' world views and lifestyles; third, *committed,* in that the geographer attempts to make his theories practically useful to real-world people and situations which in turn can provide an accurate context from which genuine description and theory can arise.

Although in the last few years there have been a host of volumes on the philosophy of geography discussing a great variety of philosophical approaches and outlooks,[5] Gregory is probably correct in highlighting a positivist mainstream in geographic research and Marxist-structural and reflexive perspectives as the two major alternative philosophical thrusts in the discipline. Gregory's argument is therefore significant because it seeks to identify present philosophical weaknesses in geography and to offer one potential direction which may lead to an integrated discipline.[6] In today's times of financial retrenchment in academia, discussion of new research directions is not entirely popular as many scholars rely on tested, conventional approaches rather than untried, less familiar methods and points of view.[7] Yet if Gregory is even partially right in suggesting that an uncritial positional geography will eventually lead to its own collapse, then his argument is important and requires discussion in all subfields of the discipline. This essay marks a beginning toward that aim, examining the value of a reflexive perspective in behavioral geography. Discussion is limited to the substantive themes that

---

5. E.g., David Ley and Marwyn Samuels, eds., *Humanistic Geography: Prospects and Problems* (Chicago: Maaroufa Press, 1978); Steven Gale and Gunnar Olsson, eds., *Philosophy in Geography* (Dordrecht, the Netherlands: D. Reidel, 1979); Milton E. Harvey and Brian P. Holly, eds., *Themes in Geographic Thought* (London: Croom Helm, 1981); David Stoddart, ed., *Geography, Ideology and Human Concern* (Totawa, New Jersey: Barnes and Noble, 1981); Peter Gould and Gunnar Olsson, eds., *A Search for Common Ground* (New York: Methuen, 1982).

6. Reaction to Gregory's argument has been various. Two of the most thoughtful discussions are Robert Sack's review in *Progress in Human Geography* 3 (September 1979): 443-452; and Richard Walker's review in *Annals of the Association of American Geographers* 69 (September 1979): 518-520.

7. For a statement indicating the overwhelming impact of a positivist outlook on conventional research funding in geography, see Barry M. Moriarty, "Future Research Directions in American Human Geography," in *Professional Geographer* 33 (1981): 484-488. For an effective counter to Moriarty's presentation, especially from a phenomenological angle, see John Pickles, "'Science' and the Funding of Human Geography," in *Professional Geographer* 34 (1982): 387-392.

reflexivity--specifically, a phenomenological approach--can offer behavioral geography. The argument is that a phenomenological perspective looks at the person-environment relationship anew and thus helps to revitalize the methodological, epistemological and ontological foundations of behavioral geography.

Aims of a Reflexive Behavioral Geography

Literally, reflexivity is a turning-back on one's self. The aim is an explicit look at points of view and ways of living that are normally implicit, unquestioned, and taken-for-granted. The recognition is that the person both creates and responds to a variety of systems of meanings, all of which need self-conscious consideration and understanding. The philosophical traditions most closely associated with reflexivity are existentialism, hermeneutics and, especially, phenomenology.[8] In geography, the reflexive perspective has most commonly been called "humanistic",[9] though some geographers have used this label in wider fashion to incorporate a Marxist-structural perspective.[10] A major aim of the reflexive perspective, especially a phenomenonological thrust, is a self-conscious consideration of *lifeworld,* the taken-for-granted pattern and context of everyday living through which the person conducts his or her day-to-day life without having to make it an object of conscious attention.[11] Immersed in their daily world of cares and concerns, people normally do not consider the lifeworld; it is concealed as a phenomenon. The phenomenologist works to unmask the lifeworld's concealment, bringing its aspects and qualities to explicit scholarly attention.

For the phenomenological geographer, a first concern is the geographical aspects of the lifeworld, or *geographical lifeworld,* as one might call it. The geographical lifeworld is the sum of taken-for-granted meanings, experiences, behaviors and events in relation to environment, space, place and landscape. The

---

8. On an existential geography, see Marwyn S. Samuels, "An Existentialist Geography," in Harvey and Holly, eds., op. cit., pp. 115-132; on a hermeneutical geography, see Robert Mugerauer, "Concerning Regional Geography as a Hermeneutical Discipline," *Geographische Zeitschrift* 69 (January 1981): 57-67; Courtice Rose, "Wilhelm Dilthey's Philosophy of Historical Understanding: A Neglected Heritage of Contemporary Humanistic Geography,: in Stoddart, ed., op. cit., pp. 99-133; on a phenomenological geography, see Edward C. Relph, "Phenomenology," in Harvey and Holly, eds., op. cit., pp. 99-114; and David Seamon, "The Phenomenological Contribution to Environmental Psychology," *Journal of Environmental Psychology* 2 (June 1982): 119-140.

9. E.g., Ley and Samuels, eds., op. cit.

10. Thomas F. Saarinen and James L. Sell, "Environmental Perception," *Progress Human Geography* 4 (December 1980): 525-548.

11. Anne Buttimer, "Grasping the Dynamism of Lifeworld," *Annals of the Association of American Geographers* 66 (June 1976): 277-292; David Seamon, *A Geography of the Lifeworld* (New York: St. Martin's Press, 1979).

geographical lifeworld clearly has links with other aspects of the lifeworld--for example, its social, economic and political dimensions--and eventually, these links must be identified. Yet a phenomenological geography begins first with the geographical lifeworld, aiming to identify its essential qualities and interconnections as thoroughly as possible.

Conventional behavioral geography, in focusing on themes like cognitive mapping, environmental preferences, and territoriality, has made major strides in examining the geographical lifeworld. The phenomenological criticisms are, first, that the approach taken has generally been positivist; second, that the *a priori* theoretical perspectives frequently used have led to an incomplete picture of people's behavioral and experiential relationship with environment.[12] Furthermore, conventional behavioral geography, again largely because of positivist requirements, has given little attention to the experiential qualities of landscape, environment and place. What, for example, are the experiential qualities that transform a physical space into a place and infuse it with *genius loci*--i.e., an atmosphere or character? How is the countryside different from the city in terms of behavior, experience, and ambience? In what ways experientially does a steppe environment differ from valley or mountainous landscapes and how do these environments have a bearing on the personalities and lifestyles of their resident groups?

Because of its emphasis on explanation and empirical measurement, positivist behavioral geography has given little attention to these kinds of questions. Phenomena such as "sense of place" or "regional personality" are first of all questionable in the positivist view because they are not readily identifiable in tangible form. Second, the positivist asks, assuming they do exist, how can such phenomena be quantitatively recorded and measured? In accepting alternative criteria for reality, identification and validity, a phenomenological perspective can more easily address questions dealing with less readily identifiable and explainable phenomena and processes. From one point of view, it can be said that a phenomenological behavioral geography explores two major themes: (1) a phenomenology of environmental behavior and experience; (2) a phenomenology of the person-environment relationship. Each theme is discussed in turn, though ultimately these topics are intimately connected and together establish the foundation for a thorough understanding of the geographical

---

12. Seamon, "Phenomenological Contribution."

lifeworld.[13]

A Phenomenology of Environmental Experience and Behavior

A first question asked by a phenomenological behavioral geography is in what ways do we as human beings reach out, make contact with, and behave in the geographical lifeworld? A key phenomenological notion relevant here is *intentionality*--the fact that all human impulses, awareness, and actions do not exist unto themselves but are directed toward something and have an object.[14] One phenomenological task is to identify the modes and range of human intentionality. The result is a presentation of human experience and behavior which is considerably more complex and rich than portraits provided by traditional philosophy and psychology.

One predominant model in conventional behavioral geography, for example, is person as cognitive being. The assumption is that people are rational decision-makers whose spatial and environmental behaviors are largely the result of cognitive processes.[15] The phenomenological criticism of a cognitive approach to environmental behavior is that the geographical lifeworld is reduced to what people think and say it is. A phenomenological perspective realizes the significance of cognitive intentionality but also recognizes other important modes of experience and behavior--e.g., emotional and bodily intentionalities.

Research on the body's role in experience and behavior has become especially significant in phenomenology because of the phenomenologist's wish to circumvent and reinterpret the Cartesian split between body and mind common in traditional Western philosophy and psychology. Reality, according to the Cartesian view, can be divided into two substances--thinking substance (mind, or consciousness) and extended substances (bodies); this division "set the stage for the problems in modern philosophy which center around the question of how these substances can have an epistemological relationship."[16] A phenomenological perspective avoids Cartesian dualism in part by arguing that there are other modes of consciousness and intentionality besides cognitive mind, including bodily modes. Phenomenological research on these modes can be

---

13. The contents of these themes can only be indicated here. For further discussion, see Seamon, "Phenomenological Contribution," on which the two following sections are based.

14. D. Stewart and A. Mikunas, *Exploring Phenomenology* (Chicago: American Library Association, 1974).

15. E.g., see the majority of essays in Kevin R. Cox and Reginald G. Golledge, eds., *Behavioral Problems in Geography Revisited* (New York: Methuen, 1981).

16. Stewart and Mikunas, op. cit., p. 4.

discussed around two interrelated themes: (1) the body-as-given, and (2) the body-as-learned.[17]

Work on the body-as-given examines how the human body's inherent structure, size and functions work to circumscribe and sustain the range of human intentionalities and actions in the world. Straus, for example, has examined the role of upright posture in organizing human encounter with the world,[18] while Dovey has considered the significance of the body's six-directional axis of up-down, left-right, and front-back in creating a sense of "symbolic space."[19] Phenomenological research also examines the way in which the five senses and perception set horizons of human experience and the resulting "sense-scapes."[20] The main aim of phenomenological work on the body-as-given is to attune people to the crucial role that bodily structure and process has in conferring on the human situation a particular order and style of existence.

Phenomenological research on the body-as-learned, in contrast, examines the precognitive but learned intentionality of the body, which after the French phenomenologist Maurice Merleau-Ponty,[21] has come to be known as *body-subject*.[22] Body-subject is the preconscious intelligence of the body manifested through action.[23] It is the bedrock of habitual, routine behaviors, including most everyday spatial behaviors, and indicates why a large portion of a person's typical day can proceed without a major amount of cognitive attention and conscious awareness. One important aim of a phenomenological behavior geography is to ask how body-subject works in extended ways over time and space, and to ask how routine behaviors of individuals coming together in space can transform that space into a place with a particular dynamism and character.[24] A

---

17. Seamon, "Phenomenological Contribution."

18. Erwin Straus, "The Upright Posture," in *Phenomenological Psychology* (New York: Basic Books, 1966), pp. 137-165.

19. Kimberly G. Dovey, "The Dwelling Experience: Towards a Phenomenology of Architecture," *University of Melbourne Department of Architecture and Building Research Paper* no. 55 (Melbourne: University of Melbourne, Department of Architecture and Building, 1978), pp. 38-58.

20. E.g., Donald Ihde, *Sense and Significance* (Pittsburgh: Duquesne University Press, 1973); Yi-Fu Tuan, *Topophilia* (Englewood Cliffs, New Jersey: Prentice-Hall, 1974); R. Murray Schafer, *The Tuning of the World* (New York: Knopf, 1977), Lisa Heschong, *Thermal Delight in Architecture* (Cambridge, Mass.: MIT Press, 1979).

21. Maurice Merleau-Ponty, *The Phenomenology of Perception* (New York: Humanities Press, 1962).

22. Mary Rose Barral, *Merleau-Ponty: The Role of Body-Subject in Interpersonal Relations* (Pittsburgh: Duquesne University Press, 1965).

23. Seamon, *Geography of the Lifeworld*, p. 41.

24. Ibid., chaps. 6 and 19, David Seamon and Christina Nordin, "Marketplace as Place Ballet: A Swedish Example," *Landscape* 24 (October 1980): 38-45.

second important aim is to consider how physical design can isolate people or bring them together in place. Here, work like that of Jane Jacobs, Oscar Newman, and William Whyte, though not directly phenomenological, is crucial, since these researchers identify clear links between design qualities and environmental satisfaction and interaction.[25]

Body-subject is a useful notion to overcome the gap between cognitive and behaviorist theories--the two major perspectives from which spatial behavior has been interpreted in conventional psychology.[26] In contrast to the stimulus-response argument of the behaviorists, the phenomenological interpretation is that the precognitive intelligence of body-subject is not a chain of discrete passive responses to external stimuli; rather the body possesses within itself an active, intentional, holistic capacity which 'knows' through action the everyday spaces and environments in which the person lives.[27] In contrast to cognitive theories, the phenomenological argument is that a sizable portion of behaviors in the lifeworld is pre-cognitive and involves the prereflective intentionality of body-subject.[28] Eventually, the phenomenological behavioral geographer must examine the dialectical relationship between the innovative, active force of cognition and the passive, conservative force of body-subject, habit, and routine. How the polarity between these two forces resolves itself in particular situations is a key to understanding why a person's environmental behaviors cannot always change, even if cognitively he or she wants them to become different.

A Phenomenology of the Person-Environment Relationship

A phenomenological behavioral geography also re-examines the person-environment relationship. Traditionally, this relationship has been described in terms of a subject-object dichotomy spoken of in terms of an *idealist* position (arguing that man acts on the world, which is organized by human consciousness) or a *realist* position (arguing that the world acts on man, who is largely the product of forces outside himself in the external environment). A phenomenological view argues against any person-world dualism:

---

25. Jane Jacobs, *The Death and Life of Great American Cities* (New York: Vintage, 1961); Oscar Newman, *Community of Interest* (New York: Doubleday, 1980); William H. Whyte, *The Social Life of Small Urban Spaces* (Washington, D.C.: Conservation Foundation, 1979).

26. Merleau-Ponty, op. cit.

27. Richard M. Zaner, *The Problems of Embodiment: Some Contributions to a Phenomenology of Body* (The Hague: Martinus Nijhoff, 1974).

28. Merleau-Ponty, op. cit.; David Seamon, *Lifeworld*, chaps. 3-7.

human consciousness and experience necessarily involve some aspect of world as their object, which in turn provides the context for the meaning of consciousness and experience. There is, in other words, "an undissolvable unity" between person and world, or *being-in-the-world*, as phenomenologists often call it to emphasize a sense of immersion and integral person-world fusion.[29]

A major phenomenological theme used to maintain the wholeness of this person-world indissoluability is the notion of *place,* which Relph defines as a fusion of human and natural order and a significant center of a person's immediate experience of the world.[30] Relph's argument is that places have meaning in direct proportion to the degree that one feels *inside* that place--i.e., here rather than there, enclosed rather than exposed, secure rather than threatened. Next, one can speak of varying degrees of insideness--for example, *empathetic insideness,* a situation where a person who is an outsider in terms of place works through concern, interest, and empathy to understand that place and come to know its essential meaning and structure; or *existential insideness,* a situation of profound, unself-conscious immersion in place and the experience most people know when they are in their own community or region. On the other hand, one can feel *outside* place, which might involve, for example, *existential outsideness*--a sense of alienation and homelessness; or *objective outsideness*--the intentional separating of person from place in order to study it selectively in terms of one particular attribute or activity.

Though clearly incipient and not inclusive, Relph's inside-outsideness continuum is an important beginning for providing a self-conscious presentation of place experience which applies to particular places yet extends beyond them to help people understand their environmental dealings in more general, reflexive terms. Relph's continuum sensitizes students to different modes of place experience, helping them to realize that the same place may foster considerably different modes of insideness and outsideness for different individuals and groups. Consider, for example, the use of this continuum in the classroom. Students are asked to select a mode of insideness or outsideness with which they feel familiar and to describe it in writing. A Nigerian student depicts the strong existential outsideness he felt when first arriving in America, while a woman who grew up in a small Kansas town pictures the intense sense of existential insideness she knew there. Yet another

---

29. Stewart and Mickunas, op. cit., p. 9.
30. Edward C. Relph, *Place and Placelessness* (London: Pion, 1976), p. 141.

student, formerly a travelling salesman, speaks about the incidental outsideness he felt passing through places that were little more than backgrounds for his primary aim of selling. Next, students are asked to assemble in small groups and describe to each other their various accounts. The result is a self-conscious awareness of experiences and places which before were unreflected upon and therefore unnoticed.

Buttimer suggests that ultimately it is the empathetic insider who is best able to mediate between people who live in places and people who wish to plan for those places.[31] The empathetic insider attempts to understand environment, place and region as they are in their own fashion and therefore takes the time and effort to grasp the insider's life and to render an accurate account. An important contribution that the phenomenological perspective can offer the student wishing to become an empathetic insider is a set of concepts portraying essential structures of place and lifeworld. Ralph's inside-outside continuum is successful in this regard, and one aim of phenomenological research on the geographical lifeworld is to provide other descriptive notions and structures that will help students become more sensitive to their environmental and place experiences.[32]

If the phenomenon of place can be studied from the focus of human experience, as in Relph's work, it can also be examined from the angle of physical environment and landscape. In this approach, place is as genuine a phenomenon as the individual human beings coming together and forming the human dimensions of that place. Place, in other words, is an environmental synergy larger than the human parts that help to comprise it. Place has a certain quality--the Norwegian architect Norberg-Schulz calls it *genius loci*, or spirit of place--which is largely unmeasurable, unobjectifiable, and therefore very much inaccessible to conventional positivist methods.[33] Norbert-Schulz goes so far as to claim that places are permanent environmental-human wholes which maintain themselves temporally as actual events and generations come and go. Places, in other words, have an integral character remaining constant over time, extending beyond the particular people living in that place at a particular historical moment. This view

---

31. Anne Buttimer, "Home, Reach, and a Sense of Place," in *The Human Experience of Space and Place,* ed. Anne Buttimer and David Seamon (London: Croom Helm, 1980).

32. David Seamon, "Phenomenology, Geography and Geographic Education," *Journal of Geography and Higher Education* 3 (Autumn 1979): 40-50.

33. Christian Norbert-Schulz, *Genius Loci: Towards a Phenomenology of Architecture* (New York: Rizzoli, 1980).

contradicts the conventional behavioral approach to environment, where place is seen to be only the sum of the various psychological, social, economic and political forces working in an environment at a particular point in time. In this conventional view, unique qualities of the place itself are secondary or ignored; the assumption is that a suitable manipulation of the right external factors can change the nature of place. Norberg-Schulz, in contrast, argues that places are essentially what they are and that human intervention will be successful when it recognizes that essence and creates human environments in tune with it rather than out of touch and therefore imposed, arbitrary and discordant. The lesson is that environments "must be treated as *individual places,* rather than as abstract spaces where the 'blind' forces of economy and politics must have free play."[34]

In his book, Norberg-Schulz asks how the qualities of natural environment meet together in place to create the place's special style and character. He establishes a fourfold typology of natural environment grounded in such qualities as spatial character, light, and temporal rhythms.[35] Norberg-Schulz's aim is to identify how atmosphere, water, land and life meet in location to generate a particular style of place and region which have a definite experiential "feel" about them. The recognition here, as in Relph's work, is that involvement with environment and place is an essential, inescapable aspect of human existence which can be clarified through qualitative description. Whether starting from the focus of human experience (as with Relph), or from the focus of geographical environment (as with Norberg-Schulz), the phenomenological recognition is that any subject-object or person-environment relationship is an intellectual illusion out of touch with experiential and existential reality. Rather, people are immersed in their world, which in part is geographical. Understanding the full range of this geographical immersion is one central aim of a phenomenological behavioral geography.

Conclusion

Much more could be said about the phenomenological contribution to behavioral geography--e.g., Heidegger's notions of dwelling and appropriation,[36] discussions on the emotional

---

34. Ibid., p. 182.

35. Ibid., chaps. 1-3.

36. Martin Heidegger, *Poetry, Language, and Thought* (New York: Harper and Row, 1971); Joseph Grange, "On the Way Towards a Foundational Ecology," *Soundings* 60 (1977): 135-149; Norberg-Schulz, op. cit.; Edward C. Relph, *Regional Landscapes and Humanistic Geography* (London: Croom Helm, 1981).

dimensions of human environmental experience and behavior,37 or research examining the sacred meaning of space, place and environment.38 This paper has emphasized the phenomenologist's wish to re-examine and extend the behavioral presentation of environmental experience and behavior and to re-interpret the person-environment relationship in a way which avoids a sense of dichotomy and dualism. This focus is only one possible way in which reflexive research can move. Ley, for example, has identified such work as "sense of place studies."39 His preference, instead, is for reflexive research that centers on human existence but which also emphasizes "a discussion of material and power relations and the examination of contexts that include, but also fall beyond, the consciousness of man...."40

Much could also be said about the problems and weaknesses of a reflexive approach in behavioral geography. Criticisms have been drawn by both positivists and Marxist-structuralists.41 Ley summarizes these complaints as threefold: first, that reflexive research often focuses on trivial or unique subject matter; second, that it frequently lacks in scientific objectivity; third, that it generally ignores the social, political, and economic structures which constrain individual freedom and lead to power relations.42 These censures are not so much flaws of the reflexive approach itself as they are criticisms which arise from the ideologies and personal preferences of the positivist (who sees empirical validation, explanation and pragmatic utility as the prime aims of academic research) or the Marxist-structuralist (who sees the prime

---

37. Yi-Fu Tuan, *Topophilia*; Seamon, *Lifeworld*.

38. Mircea Eliade, *The Sacred and the Profane* (New York: Harcourt, Brace and World, 1957); Seyyed Hossein Nasr, "Foreword," in Nadar Ardalan and Laleh Bakhtiar, *The Sense of Unity: The Sufi Tradition in Persian Architecture* (Chicago: University of Chicago Press, 1974): xi-xv; Diana L. Eck, *Banaras: City of Light* (New York: Knopf, 1982).

39. David Ley, "Behavioral Geography and the Philosophies of Meaning," in *Behavioral Problems*, ed. Cox and Golledge, p. 219.

40. Ibid., p. 225.

41. Positivist critique includes Leslie J. King, "Alternatives to a Positivist Economic Geography," *Annals of the Association of American Geographers* 6 (June 1976): 293-308. Marxist-structural critique includes Dennis Cosgrove, "Place, Landscape, and the Dialectics of Cultural Geography," *Canadian Geographer* 22 (Spring 1978): 66-72; Kevin R. Cox, "Bourgeois Thought and the Behavioral Geography Debate," in *Behavioral Problems*, ed. Kevin R. Cox and Reginald G. Golledge, pp. 256-279. Especially interesting is the exchange of essays between Gregory and Ley: Derek Gregory, "Human Agency and Human Geography," *Transactions, Institute of British Geographers* 6 (1981): 1-18; David Ley, "Rediscovering Man's Place," *Transactions, Institute of British Geographers* 7 (1982): 248-253; Derek Gregory, "A Realist Construction of the Social," *Transactions, Institute of British Geographers* 7 (1982): 254-256. A reflexive critique of the Marxist-structural perspective is provided by James Duncan and David Ley, "Structural Marxism and Human Geography," *Annals of the Association of American Geographers* 72 (March 1982): 30-59.

42. Ley, "Behavioral Geography," pp. 222-225.

research aim as the identification and change of underlying societal structures and processes that promote economic and political inequities).

In a recent commentary, Ley argues that the aim of geography in the 1980s should be "a synthesis which will incorporate both the symbolic and structural, both the realm of constraints and the realm of meanings, where values and consciousness are seen as embedded in their contexts, and where environments are treated as contingent before emerging forms of human creativity."[43] The great practical difficulty here is balancing personal preferences and ideologies. Most scholars of a reflexive bent see individual awareness and action as the significant initiator of personal and societal change, while most Marxist-structuralists place the power with external socio-economic constraints and structures, many of which can not be seen or modified directly. Yet again, the positivist hope for the future is the improvement in material well-being provided by science and technology. Even though Ley (as a major figure representing reflexivity) and Gregory (as a major Marxist-structuralist) both call out for synthetic research, philosophical penchants could well overwhelm constructive dialogue and harmonization. Geographical research, including behavioral work, may continue to be divided by philosophical and ideological preferences.

---

43. David Ley, *Geography Without Man: A Humanistic Critique,* University of Oxford School of Geography Research Paper, no. 24 (Oxford, University of Oxford, Department of Geography, 1980); cited in David Ley, "Rediscovering Man's Place," p. 249.

CHAPTER 12

POSITIVIST PHILOSOPHY AND RESEARCH ON HUMAN SPATIAL BEHAVIOR

Reginald G. Golledge and Helen Couclelis

Analytical research in geography, including a good part of the research on human spatial behavior, owes much to its positivist roots. The introduction of positivism in geography was a critical part of the quantitative and theoretical revolution that swept the discipline in the late 1950s and early 1960s. The combination of positivist philosophy, quantitative methods and normative theory pushed the discipline as a whole into a new era of scientific thought.

It soon became obvious that, while some of the fundamental principles of positivism are ingrained in most if not all analytical work, the limitations and constraints of the philosophy itself were too confining for much geographic research. For example, positivism insists on a physicalist view of the flux of being; the characterization of reality as a collection of atomistic facts; an emphasis on the observable; the critical nature of hypothesis testing within the context of a particular mode of reasoning called "scientific method;" an empiricist base; a search for generalizations; aspiration towards producing process theory; a need for public verification of results; the necessity of logical thinking; the separation of value and fact; the scientist as a passive observer of an objective reality; and the principle that value judgments must be excluded from science.

While a substantial part of behavioral research in geography was born in the positivist tradition and still reflects some of the underlying principles, recent analytical research has progressed so far beyond that original basis that it has become increasingly difficult to identify such research with traditional positivism. In particular, the analytical research of the 1970s and 1980s has embraced a wider epistemological base while concurrently recognizing that fewer and fewer of the classical positivist tenets now seem necessary. As we have pointed out elsewhere: "Looking back at this past decade of development, however, it is quite remarkable to see to what extent behavioral research in geography seems to have moved beyond the original positivist assumptions, while gaining momentum as an articulate and coherent perspective on the issue of human spatial behavior."[1]

Analytical behavioral research in geography has in fact evolved by shedding, one after another, various classic positivist tenets. First to go, as the importance of perceptual and cognitive images increased, was the critical tenet of the non-existence of the unobservable. With the loss of this basis came a weakening of reductionism and the physicalist interpretations of human behavior. Thus, as Cox and Golledge pointed out, spatial behavior became differentiated from behavior in space.[2] Another development forcing researchers out of a positivist mold was the adoption of transactional or constructivist positions for examining the relation between humans and the environments in which they lived.[3] This destroyed the tenet of the scientist as a passive observer of an objective reality. The transactional-constructivist position, based as it was on the dynamics of interaction between person and environment, also hastened the demise of the positivist positions of objectivism, the non-scientific importance of values and beliefs, and the separation of value and fact. Current analytical research on human spatial behavior accepts that "values" must *be* "facts," since they causally determine the facts of overt spatial behavior. An essential part of analytical behavioral research is the assumption that mind and world are in constant dynamic interaction--a far cry from the positivist position of an *a priori* world.

Analytical behavioral research over the last decade has progressed by leaping the barriers imposed by positivist philosophy while retaining those principles of scientific research that have endured the rise and fall of many philosophies. What is left, therefore, is what is truly positive in positivist thought: the importance of logico-mathematical thinking; the desire for public verifiability via intersubjective tests of knowledge by continuous

---

1. H. Couclelis and R. G. Golledge, "Analytic Research, Positivism, and Behavioral Geography," (unpublished manuscript, Department of Geography, University of California, Santa Barbara, 1982), p. 3.

2. K. R. Cox and R. G. Golledge, *Behavioral Problems in Geography*, Northwestern University Department of Geography Research Papers, no. 17 (Evanston: Northwestern University, Department of Geography, 1969), p. 17.

3. G. T. Moore and R. G. Golledge, eds., *Environmental Knowing* (Stroudsburg, Pa.: Dowden, Hutchinson and Ross, 1976).

reference to experience; a search for generalizations; and the emphasis on analytic languages for expressing knowledge.

The trend away from classic positivism has accelerated over the last decade. But this has not meant a flight to humanist alternatives. Rather, it has signified the development of an analytic mode of discourse that serves well the original aims of behavioral research in the discipline. To illustrate the essential nature of this tie, we examine the origins and purposes of behavioral research in geography, then speculate on the future role of analytic behavioral research in the discipline.

Origins and Purposes of Behavioral Research in Geography

The origins of behavioral research in human geography can be traced in part to a general movement in the social and behavioral sciences that reached its peak in the early 1960s. This was a movement toward interdisciplinary cooperation: a movement that saw huge cracks appearing in the somewhat inflexible disciplinary boundaries that developed around many academic areas by the 1950s. This urge for cross-disciplinary action saw the development of many "institutes" focusing on social, urban, behavioral, legal, environmental, and health and welfare problem areas. This research was partly a function of a widespread desire to reach deeper levels of understanding of problem situations than had been possible within the strict disciplinary context, and partly because many researchers observed that progress was being made in other disciplines toward the solution of problems central to their own areas of interest. Consequently, there arose a desire to search the epistemologies, philosophies, theories, methods and concepts of other disciplines for relevant factors that could help in the search for explanation and understanding. As part of this process, the scientist left his laboratory and entered the real world, while the real-world analyst delved into the relevance of laboratory methods. For a while, it looked as if some disciplines would never intersect. For example, psychologists in their search for information about external physical, socio-cultural, and other human environments; and geographers leaving the macro-scale studies of overt spatial activity in an attempt to examine the internal processes associated with that human activity; looked like they were going to bypass each other. Luckily this did not occur, and by the late 1960s a number of new interdisciplinary meeting places emerged where exchanges of ideas and methods checked these headlong flights and helped mold together different approaches. A classic example of this was the development of the Environmental Design and Research Association (EDRA) in which geographers, psychologists, sociologists, anthropologists, planners, designers and engineers constructively interacted with each other in an attempt to understand the roles of people and nature in the processes that were reshaping the earth.

In geography, this search took the form of investigating the relevance of themes such as image and imagination,[4] attitude and expectation,[5] risk and uncertainty,[6] decision and choice,[7] and learning and habit.[8] One of the earliest of these groups were those researchers interested in "hazard perception." This group was spearheaded by Gilbert White, Robert Kates, and Ian Burton of Chicago, who examined attitudes towards and behaviors relating to major environmental problems.[9] Others, including David Lowenthal, Yi-Fu Tuan, Anne Buttimer, and Tom Saarinen, searched for meaning in the landscape and found it best comprehended in terms of tastes and preferences, ideals and visions, values and beliefs.[10] Some researchers sought explanation and understanding through concepts of individual and environmental personalities.[11] Partly as a reaction against the levels of satisfaction attained by using the principles of optimality and equilibrium and expressing them in normative models and theories, researchers such as Wolpert, Gould, Rushton, and Golledge focused on characteristics such as

---

4. J. K. L. Wright, "Terrae Incognitae: The Place of Imagination in Geography," *Annals of the Association of American Geographers* 37 (1947): 1-15; D. Lowenthal, "Geography, Experience and Imagination: Towards a Geographical Epistemology," *Annals of the Association of American Geographers* 51 (1961): 241-260; W. Kirk, "Historical Geography and the Concept of the Behavioral Environment," *Indian Geographical Journal*: *Silver Jubilee Edition* (1951): 152-160.

5. J. Sonnenfeld, "Environmental Perception and Adaption Level in the Arctic," in *Environmental Perception and Behavior,* ed. D. Lowenthal, Chicago Department of Geography Research Paper no. 109 (Chicago: University of Chicago, Department of Geography, 1967): 42-57; R. W. Kates and J. F. Wohlwill, "Man's Response to the Physical Environment," *Journal of Social Issues* 22 (1966): 116-126.

6. P. R. Gould, "Man Against His Environment: A Game Theoretic Framework," *Annals of the Association of American Geographers* 53 (1963): 290-297.

7. J. Wolpert, "The Decision Process in a Spatial Context," *Annals of the Association of American Geographers* 54 (1964): 536-558; J. Wolpert, "Behavioral Aspects of the Decision to Migrate," *Papers and Proceedings of the Regional Science Association* 15 (1965): 159-169.

8. R. G. Golledge, "Conceptualizing the Market Decision Process," *Journal of Regional Science* 7 (1967): 239-258.

9. Examples of research in "environmental perception" are summarized in T. Saarinen, *Environmental Planning*: *Perception and Behavior* (Boston: Houghton Mifflin Company, 1976).

10. Typical papers include: A. Buttimer, *Values in Geography,* College Geography Resource Paper no. 24 (Washington, D. C.: Association of American Geographers, 1974); Yi-Fu Tuan, "Humanistic Geography," *Annals of the Association of American Geographers* 66 (1976): 266-276; D. Lowenthal, "Geography, Experience and Imagination: Towards a Geographical Epistemology," *Annals of the Association of American Geographers* 51 (1961): 241-260.

11. J. Sonnenfeld, "Personality and Behavior in Environment," *Annals of the Association of American Geographers* (1969): 136-140.

decision-making and choice behavior, risk and uncertainty, learning and habit, preference and choice.[12] Following Gould's attempts[13] to represent peoples' preferences for places in map-like form ("mental maps"), Downs[14] delved into psychiatric theory to bring to geography a model and a method for recovering cognitive images of segments of the built environment.[15] While these lines of research were followed in the United States and the United Kingdom, earlier research by Hagerstrand and his students in Sweden (carried to the English speaking world by researchers such as Pred and L. Brown), focused on the tremendous importance of search for, and flow of, information and the processes of diffusion and adoption.[16]

All of these research activities represented a desire to increase the geographer's level of understanding of particular types of problems. In search for additional concepts, additional theories, additional models, and additional relevant variables, the profession as a whole was exposed to a mass of interdisciplinary material that was both strange and new.

Perhaps the single most important characteristic for the human geographer generally was the change in emphasis from form to process. The message of the behavioral research was clear: superficial description of what is there (whether verbally or mathematically) was no longer enough. What was required was insight into why things were there. Thus, within the space of a decade, this segment of research in human geography changed from one that was interested primarily in classification and categorization of phenomena on the earth's surface, in which a particular device for classification (the region) dominated and in which a particular procedure (areal association) was a major analytical tool, through a theoretical and quantitative revolution that sought to find where things "ought to be" by establishing norms or ideal patterns that were abstractions from reality. Whether this search for explanation was undertaken by means of comprehensive description, by abstract modeling, or by the use of languages such as mathematics and

---

12. Representative samples of this work include J. M. Blaut and D. Stea, "Studies of Geographic Learning," *Annals of the Association Geographers* 61 (1971): 387-393; R. G. Golledge, "Conceptualizing the Market Decision Process."

13. P. R. Gould, "On Mental Maps," in *Image and Environment* ed. R. M. Downs and D. Stea (Chicago: Aldine, 1973), pp. 182-220.

14. R. Downs, "The Cognitive Structure of an Urban Shopping Center," *Environment and Behavior* 2 (1970): 13-39.

15. G. A. Kelley, *The Psychology of Personal Constructs* (New York: Norton, 1955).

16. Some of Hagerstrand's early work on migration and diffusion is reviewed in A. Pred, *Behavior and Location* (Lund, Sweden: Lund Studies in Geography, Gleerup, 1966).

statistics, it became obvious that research in human geography had received a shock from which it might never recover and which consequently has played a major role in the events of the discipline for more than two decades.

Apart from being primarily process-driven, behavioral research can be said to have the following purposes:

1. To change the focus of research from the artificial aggregate levels that dominated much geographic research to the disaggregate individual and group level.

2. To provide alternative models of man, especially alternatives to the omniscient economically rational or spatially rational being that of necessity was incorporated into normative theories and models.

3. To search for new models of "environment" other than the observable external physical environment. These models included not only the well known economic, social, political and cultural environments, but also perceptual, cognitive, ideological, philosophical, psychological, and other subjective environments that were part of the overall dialectic between humans and the world outside their minds and bodies.

4. To increase the set of relevant variables from which geographers could draw their explanatory schema by searching associated or allied disciplines. This search was accompanied by a rejection of the relevance of existing data banks and a move to create new data sets by survey research and other interactive methods. This meant an emphasis on new methods of data analysis (including non-parametric analytic measures and complex experimental designs) and a resulting need to explore the relevance of multi-dimensional methods of representation and analysis of phenomena.

5. To indicate to other disciplines that some of the concepts, theories, and models central to their perspectives frequently had an important spatial component that had heretofore been neglected and that should be incorporated into their own explanatory schema.

6. To define, then to attempt to represent externally, those internal processes that were critical in producing the final form of interactions between humans and the physical environment (i.e., to discover what was inside what the system's analyst called the "black boxes" interspersed between man and environment). This also implied the search for new languages in which to express survey findings, and the development of new variables relating to the internal

processes that dominate human thought and action (e.g., learning, perception, thinking, attitude formation, values, beliefs, images, emotions, personalities, and many other relevant characteristics of use in helping to understand relations between humans and their various environments).

7. To illustrate that generalizations in geography had often been made from the wrong point of view. For example, it was pointed out that generalizations made on the basis of analyzing a census reflected more the structure of the census than the behavior of people. Thus, a search was begun for a new basis for generalization by attempting to find sets of behaviors, internal or external, about which researchers could make legitimate, more general, more realistic and appropriate statements about human actions. It was hoped that if new theory were to be produced in geography, this new basis for generalization would provide a more appropriate building block on which to build those theories.

Analytic Thought in Behavioral Geography

As can be seen from the above discussion of the origins and purposes of behavioral research in geography, many of the criticisms that have been leveled at this type of research activity are ill-informed.[17] Even the analytic research in this area has not been largely dominated by positivist thought; it was not dominated by mathematical modelers nor by principles of objectification; and it was not represented by Skinnerian behaviorists who wanted to modify human behavior.[18] But neither was it completely dominated by non-theoretical, non-quantitative, non-analytic, non-model oriented, or anti-mathematical types of researchers. Neither positivism nor phenomenology nor any other doctrine can be said to be the philosophy underlying behavioral research in geography. To attribute all behavioral research to one or the other philosophical base and the accompanying sets of approaches is perhaps the epitome of ignorance with respect to research and thought in this area.

Just as it is obvious that no single epistemology or philosophy dominated the processes or reasoning and thought with respect to behavioral research in geography, so is it obvious that in particular research areas, one or another epistemology did tend

---

17. Cf. M. J. Boyle and M. E. Robinson, "Cognitive Mapping and Understanding," in *Geography in the Urban Environment*, ed. D. H. Herbert and R. J. Johnston (New York: Wiley, 1979), pp. 59-82; T. E. Bunting and L. Guelke, "Behavioral and Perception Geography: A Critical Appraisal," *Annals of the Association of American Geographers* 69 (1979): 448-462; E. Relph, "Seeing, Thinking, and Describing Landscapes," paper presented at the national meetings of the Association of American Geographers, San Antonio, Texas, April, 1982; and published as chapter 14 in this volume.

18. E. Relph. loc. cit.

to be emphasized more than others. Thus, for those interested in landscape description and identifying characteristics of places, various humanist philosophies such as phenomenology, idealism, and existentialism appeared to be appropriate. In other areas where attempts were made to represent externally the internal processes of thinking human beings, a range of scientific approaches appeared to dominate, such as logical positivism, reductionism, materialism, instrumentalism, and so on.

Research dominated by scientific approaches appears widely in the study of cognitive mapping and spatial behavior; attitudes and behavior; utility, choice, preference and behavior; search and learning behavior; mode choice and travel behavior; and mobility and migration behavior. Each of these areas appears to lend itself more to the processes of measurement, model building, abstraction, generalization, and theory building than other areas of behavioral research. This is in part because these themes are all topics for which existing theories and models serve as a base for on-going research activity which may be aimed at verifying existing theories and models, constructively criticizing and consequently altering them, or generating new thought on the role of cognitive processes in explanations of overt spatial activity. In addition, each of these areas lends itself to objective representation and measurement. For example, it is possible for a researcher to recover information from people in such a way that it allows one to represent externally segments of their environmental knowledge. One such representational form frequently chosen by geographers is the map. Once such a map is constructed by a researcher, both standard and innovative methods of analysis can be applied to the map to uncover such things as its underlying structure;[19] the best metric to use in map construction;[20] the dimensionality of features represented on the map;[21] the distortions of irregularities that appear in the map;[22] or the folds, holes, cracks or tears that

---

19. R. G. Golledge, J. N. Rayner, and V. Rivizzigno, "Comparing Objective and Cognitive Representations of Environmental Cues," in R. G. Golledge and J. N. Rayner, eds., *Proximity and Preference*: *Problems in the Multidimensional Analysis of Large Data Sets* (Minneapolis, Minnesota: University of Minnesota Press, 1982), pp. 233-266.

20. W. Tobler, "The Geometry of Mental Maps," in R. G. Golledge and G. Rushton, eds., *Spatial Choice and Spatial Behavior* (Columbus, Ohio: Ohio State University Press, 1976); G. D. Richardson, "The Appropriateness of Using Various Metrics for Representing Cognitive Maps Produced by Nonmetric Multidimensional Scaling," (unpublished master's thesis, University of California, Santa Barbara,1979).

21. V. L. Rivizzigno, "Individual Differences in the Cognitive Structuring of an Urban Area," (unpublished doctoral dissertation, Ohio State University, 1976).

22. N. Gale, "On the Nature of Distortion and Fuzziness in Cognitive Maps," (unpublished master's thesis, University of California, Santa Barbara, 1980).

should be represented in the map on the basis of the recovered information.23 Similarly, characteristics such as preferences and attitudes have been capable of measurement via uni- and multi-dimensional scaling methods for some considerable period of time.24 While recognizing the continuing controversy over the causal link between preference and choice, and attitudes and behaviors, it is also possible to represent preference and attitudinal data in cartographic format and to focus on their spatial composition.25

The literature on utility, choice, and preference has also been dominated by attempts to develop a suitable index expressing the magnitude of utility or preference which can then be used in a model format to try to replicate or predict choice processes.26 Similarly, researchers interested in search activities and the focusing or convergence of search activities toward repetitive or habitual behavior have substantial existing theories and models to draw on that can be examined for spatial relevance and thereafter adhered to, modified, or replaced according to their relevance for spatial problems.27

Of all research areas in which behavioral methods have been stressed, perhaps the greatest advances have been made in terms of the non-spatial area of mode choice and the spatial area of travel behavior.28 These areas have also been complemented by an on-going

---

23. R. G. Golledge and L. J. Hubert, "Some Comments on Non-Euclidean Mental Maps," *Environment and Planning A* 41 (1982): 169-204.

24. W. S. Torgenson, *Theory and Methods of Scaling* (New York: Wiley, 1958); D. Lowenthal, "Assumptions Behind Public Attitudes," in *Environment Quality in a Growing Economy,* ed. H. Jarrett (Baltimore, MD.: Johns Hopkins Press, 1966), pp. 128-137; R. Downs, "The Cognitive Structure of an Urban Shopping Center," *Environment and Behavior* 2 (1970): 13-39; G. Rushton, "Analysis of Spatial Behavior by Revealed Space Preferences," *Annals of the Association of American Geographers* 59 (1969): 391-400; T. F. Saarinen and R. U. Cooke, "Public Perception of Environmental Quality in Tucson, Arizona," *Journal of the Arizona Academy of Sciences* (1971): 260-274.

25. P. R. Gould and R. R. White, *Mental Maps* (Harmonsworth, England: Penguin, 1974); K. R. Cox and G. Zannaras, "Designative and Appraisive Characteristics of Macro Spaces," in *Images and Environment,* ed. R. M. Downs and D. Stea (Chicago: Aldine, 1973); J. Louviere and R. Meyer, "A Model for Residential Impression Formation," *Geographical Analysis* 8 (1976): 479-486; T. F. Saarinen, "The Use of Projective Techniques in Geographic Research," in *Environment and Cognition,* ed. W. H. Ittleson (New York: Seminar Press, 1973), pp. 29-52.

26. T. R. Smith, "Uncertainty, Diversification and Mental Maps in Spatial Choice Problems," *Geographical Analysis* 10 (1978): 120-141; T. Smith, W. A. V. Clark, J. Huff and P. Shapiro, "A Decision Making and Search Model for Intraurban Migration," *Geographical Analysis* 11 (1979): 1-22; W. A. V. Clark, "Recent Research on Migration and Mobility: A Review and Interpretation," *Progress in Planning* 18 (1982): 1-56.

27. C. H. P. Schneider, "Models of Space Searching in Urban Areas," *Geographical Analysis* 7 (1975): 173-185; Peter Rogerson, "Spatial Models of Search," *Geographical Analysis* 14 (1982): 217-228; P. Rogerson and R. MacKinnon, "A Geographical Model of Job Migration and Unemployment," *Papers of the Regional Science Association* 48 (1982): 89-102.

28. L. P. Burnett, "The Dimensions of Alternatives in Spatial Choice Process," *Geographical Analysis* 5 (1973): 181-204; N. Wrigley, "Categorical Data Analysis," in *Quantitative Geography,* ed. N. Wrigley and R. J. Bennett (London: Routledge and Kegan Paul, 1981), pp. 111-122.

concern for understanding the structure of decision processes, and much of this research is focused in the area of mobility within urban areas.29 The recent convergence of time-path analysis and disaggregate movement behavior is another example of the expansion of an analytic mode of discourse.30

There are in existence now a variety of well documented books and papers which summarize the specific research activities in each of the above areas and which emphasize both the epistemological bases of the research and the reasons for the choice of analytic or scientific approaches to problem solving.31 We do not, therefore, plan to spend more time on summarizing the extensive nature of this work. The overall message, however, is simple. We believe that research in these areas has adopted an appropriate philosophical base and an appropriate method of attack and that the results of this research more than justify such choices.

Future Research on Human Spatial Behavior

As we pointed out in the beginning of this paper, much of the research undertaken within the field of geography that adopts a behavioral approach can no longer be described as classically positivistic. Some general principles of positivism (such as the search for process knowledge, a search for generalizability, an empirical emphasis, and the selection of scientific modes of reasoning and analysis) are widespread and will undoubtedly continue to be used. Adoption of the analytic mode of thought brought geographers to a stage where they could more freely interact with and comprehend researchers in a large number of other disciplines. This cross-disciplinary interaction has been extremely fertile in the search for solutions to some problems and in terms of the general rate of knowledge accumulation. For example, once the geographer had brought to bear all his or her experience and skills in cartographic analysis on the problems of representing and analyzing cognitive information about places, it was possible to produce explicit theories of the process of learning about spatial

---

29. R. G. Golledge, "A Behavioral View of Mobility and Migration Research," *The Professional Geographer* 32 (1980): 14-21; W. A. V. Clark, "Recent Research on Migration and Mobility: A Review of Interpretation," *Progress in Planning* 18 (1982): 1-56.

30. A. Pred, "The Choreography of Existence: Comments on Hagerstrand's Time Geography," *Economic Geography* 53 (1977): 207-221; A. Pred, "Of Paths and Projects: Individual Behavior and Its Societal Context," in *Behavioral Problems in Geography Revisited,* ed. K. Cox and R. Golledge (London: Methuen, 1982), pp. 231-255; T. Hagerstrand, "What About People in Regional Science," *Papers of Regional Science Association* 24 (1970): 7-21.

31. Examples can be found in such recent books as K. R. Cox and R. G. Golledge, eds., op. cit.; R. M. Downs and D. Stea, eds., *Maps in Minds* (New York: Harper and Row, 1977); S. Gale and G. Olsson, eds., *Philosophy in Geography* (Dordrecht: Reidel, 1979); G. T. Moore and R. G. Golledge, eds., op. cit.

phenomena and to begin the process of modelling human decision structures.[32] These steps appear to have been particularly important to recently developing areas such as artificial-intelligence modelling. For example, three experimenters are currently conducting an examination of how children acquire and use spatial information during way-finding tasks in large-scale suburban environments.[33] These researchers will be testing two alternative theories of the accumulation of spatial knowledge--the anchor point theory which developed in geography,[34] and an alternative version in psychology.[35] Parallel to this investigation will be the construction of a computational process model (implemented as an operational computer program) that will focus on the manner in which children's knowledge structures are organized, accessed, and modified during the performance of spatial tasks. The final task will be to try to estimate how the knowledge structure effects spatial behavior. This is but one example of how our process-oriented behavioral research has contributed to the search for knowledge in an interdisciplinary context, these disciplines including computer science, psychology, geography, and the emerging field of cognitive science.

It should be pointed out, however, that activities such as these are the result of a long-term commitment to behavioral research practices. We wish to stress that on-going behavioral research using a scientific approach is not an easy thing. There are immense difficulties involved in terms of experimental design, problem formulation, data collection, data analysis, methodology selection, relation to theory, verifiability, reliability, and so on, all of which take considerable time and effort. The behavioral researcher who is interested in using a scientific approach needs to be committed to many years of study even for small problems. One should not expect to be able to define a problem of any significance

---

32. J. M. Blaut and D. Stea, "Studies of Geographic Learning," *Annals of the Association of American Geographers* 61 (1971): 387-393; R. G. Golledge and G. Zannaras, "Cognitive Approaches to the Analysis of Human Spatial Behavior," in *Environment and Cognition,* ed. W. H. Ittleson (New York: Seminar Press, 1973), pp. 59-94; R. G. Golledge, "Representing, Interpreting and Using Cognized Environments," *Papers of the Regional Science Association* 41 (1978): 169-204; R. G. Golledge and A. Spector, "Comprehending the Urban Environment: Theory and Practice," *Geographical Analysis* 10 (1978): 403-426.

33. This project is "Children's Acquisition of Spatial Knowledge of a Large-Scale Environment," by geographers T. R. Smith and R. G. Golledge, and psychologist, J. Pellegrino, all of the University of California, Santa Barbara.

34. R. G. Golledge, "On Determining Cognitive Configurations of a City," (Columbus, Ohio: Department of Geography and Ohio State University Research Foundation, 1975); R. G. Golledge, "Representing, Interpreting and Using Cognized Environments," *Papers of the Regional Science Association* 41 (1978): 169-204.

35. A. W. Siegel and S. H. White, "The Development of Spatial Representations of Large-Scale Environments," in *Advances in Child Development and Behavior,* ed. H. W. Reese (New York: Academic Press, 1975), vol. 10.

and hope to investigate or perhaps solve such a problem in the short
time spans with which many traditional geographic problems are
investigated. Undertaking scientifically grounded behavioral
research requires a long-term commitment with no guarantee of
successful results. While we are convinced that this is a viable
and, in fact, necessary procedure if we wish to advance seriously
the frontiers of behavioral research in geography, we stress that it
is not an area that should be chosen by the poorly trained, the
faint-hearted, or the dilettante. We are convinced that there is
sound opportunity as a result of adopting this approach to
contribute to theory, to develop models, and to advance the general
state of knowledge on a variety of extremely relevant behavioral
problems.

CHAPTER 13
WHODUNIT? STRUCTURE AND SUBJECTIVITY IN BEHAVIORAL GEOGRAPHY

Douglas Greenberg

This essay begins from the premise that the primary task of human geography is to explain why and how people alter, adapt to, and perceive their environments. In seeking to identify the motivating forces behind the ongoing modification of landscapes, the geographer in one sense functions as an academic detective engaged in the continuous unravelling of a geographical "whodunit?" At a superficial level, the solution might be said to be "elementary." It is, of course, the purposive, goal-oriented activity of human individuals that animates the historical stage and transforms the surface of the earth. Hence, the rationale behind much of what has come to be known as "behavioral geography" is surely sound: only when the geographer understands the intentions, cognitive processes, and actual lived experience of people can he or she claim to have an adequate grasp of human geographical phenomena.

It will be argued here, however, that since human life is intrinsically *social,* there are limitations to an excessively narrow emphasis upon the consciousness of individuals. Viewed in this light, the above geographical "whodunit?" is not the open-and-shut case it first appeared to be. I maintain in this essay that the focus upon human agency which characterizes behavioral and perceptual approaches in geography must be complemented by some conception of social *structure.* It is proposed that promising leads have been provided by that master sleuth of social theory, Karl Marx, and by some of his intellectual descendants. Overall, what I defend here is a "structuralist" reading of Marx, in which causal primacy is attributed to the generative properties of the prevailing mode of production, rather than to the consciousness or even the behavior of particular individuals.

Behavioral Geography and the Problem of the "Subject"

In general, two major approaches to behavioral-perceptual research can be distinguished: scientific-positivist and phenomenological-humanist.[1] Clearly, there are significant

---

1. See Thomas F. Saarinen and James L. Sell, "Environmental Perception," *Progress in Human Geography* 4 (December 1980): 525-548.

differences in philosophy and method separating these two broad schools of thought, but their respective adherents are united in believing that, in Roger Downs' words, "We must understand the ways in which human beings come to know and to understand the geographical world in which they live," and that "such understanding is best approached from the level of the individual human being."[2] What is maintained in common is a commitment to an individualist social ontology, and even to methodological individualism.[3] These allegiances are implicit within the conceptual schema developed for behavioral-perceptual research by Downs, which has been adapted by Thomas Saarinen for use in his popular text.[4] Based upon an essential distinction between the individual and the environment, this framework embodies the fundamental assumptions underlying most behavioral and perceptual approaches within contemporary human geography.

A major weakness of this focus upon the individual, however, is that it poses a barrier to the understanding of macro-level spatial phenomena. That the collective behavior of individuals can give rise to emergent, unintended outcomes has been acknowledged within modes of social analysis as diverse as sociological functionalism and the Hegelian philosophy of history.[5] To be sure, no fully satifactory framework for the explanation of collective phenomena has yet been developed. From business cycles to the character of urban landscapes, however, the emergent effects of social interaction continue to resist attempts to reduce them to a simple *aggregation* of the actions of individuals.[6]

---

2. Roger M. Downs, a review of John Gold, *An Introduction to Behavioural Geography,* in *Annals of the Association of American Geographers* 71 (June 1981): 319.

3. Methodological individualism is defined by one of its leading proponents as the principle which "states that social processes and events should be explained by being deduced from (a) principles governing the behavior of participating individuals and (b) descriptions of their situations," J.W.N. Watkins, "Ideal Types and Historical Explanation," in *The Philosophy of Social Explanation,* ed. Alan Ryan (London: Oxford University Press, 1973), p. 88.

4. Roger M. Downs, "Geographic Space Perception: Past Approaches and Future Prospects," *Progress in Geography* 2 (1970): 85; Thomas F. Saarinen, *Environmental Planning*: *Perception and Behavior* (Boston: Houghton Mifflin, 1976), pp. 10-11.

5. Robert Merton, "Manifest and Latent Fuctions," in *Social Theory and Social Structure* (Glencoe: The Free Press, 1957); Georg W. F. Hegel, *The Philosophy of History* (New York: Dover Books, 1956).

6. K. Knorr-Cetina, "The Micro-Sociological Challenge of Macro-Sociology: Towards a Reconstruction of Social Theory and Methodology," in *Advances in Social Theory and Methodology*: *Toward an Integration of Micro and Macro Sociologies,* ed. K. Knorr-Cetina and A. V. Cicourel (Boston: Routledge and Kegan Paul, 1981), pp. 1-47; Maurice Mandelbaum, "Societal Facts," *British Journal of Sociology* 6 (1955): 305-317.

Justifiably, behavioral and perceptual geographers have rejected the normative neoclassical economic framework which had informed previous models of spatial organization. They have had little to contribute in its stead, however, regarding the so-called "micro-macro problem." Rather, they have attempted to build theory inductively, beginning at the levels of the individual. Within this perspective, social outcomes can be treated only as the "sum of parts," simply the behavior of individuals writ large. In this regard, John Gold acknowledges in his recent text the danger of psychologism, "the fallacy in which social phenomena are explained purely in terms of facts and doctrines about mental characteristics of individuals."[7] By adding that behavioral geography should "be taken to complement, rather than compete with, the approaches adopted by, say, social, cultural, or political geographers," Gold suggests that cognitive research alone cannot be expected to provide more than a few clues toward the resolution of our geographical "whodunit?"[8]

For the most part, the intention of behavioral-perceptual geographers has been not to explain the spatial organization of society, but to illuminate the spatial behavior of individuals. Even these efforts, however, have met with only limited success. As Bunting and Guelke argue, behavioral-perceptual researchers have presumed erroneously that such elusive phenomena as "mental images" can be measured accurately.[9] Moreover, this research has failed to discover ways to link cognition to real-world behavior. Perhaps the most fundamental difficulty with cognitive approaches, however, has been their atomistic conception of the human being, which results from considering individuals in virtual isolation from their social context.

In positivist research, society is represented either as a source of "information," or else as a set of external constraints which impinge upon the individual. Overall, as Allan Pred notes, singular aspects of perception or behavior are analyzed not only as "totally divorced and unrelated to any past actions or experiences" of the people involved, but as "completely uninfluenced by the particular workings of institutions and society in general."[10] In

---

7. John R. Gold, *An Introduction to Behavioral Geography* (New York: Oxford University Press, 1980), p. 4.

8. Ibid, p. 5.

9. Trudi E. Bunting and Leonard Guelke, "Behavioral and Perception Geography: A Critical Appraisal," *Annals of the Association of American Geographers* 69 (September 1979): 455-456.

10. Allan Pred, "Of Paths and Projects: Individual Behavior and Its Societal Context," in *Behavioral Problems in Geography Revisited*, ed. Reginald G.

humanist writings, people are viewed as whole and reflective; the social, however, is reduced largely to a set of shared values or meanings, the origins of which are not explicated. If social *relations* are examined at all, they are often mistakenly equated with "intersubjectivity." Despite this failure to account adequately for humanity's social nature, geographers continue to posit the constitutive, self-contained individual as their primary object of inquiry. This may be in part because such a focus is consistent with what Clifford Geertz has identified as the distinctively Western conception of the person "as a bounded, unique, more or less integrated motivational and cognitive universe."[11]

Only recently has this dominant fixation upon the conscious, individual "subject" within philosophy and the social sciences been challenged. One major alternative to this egocentric view has become known as "structuralism." As happens so frequently with academic terms, the label "structuralist" has been applied to so many different modes of discourse that it no longer has any precise meaning.[12] Contemporary structuralism, however, is rooted mainly in the linguistic theory of Ferdinand de Saussure, in which the meaning or value of any particular unit, whether phonemic or semantic, is dependent upon its relationship with all other such units.[13] The diverse set of approaches to social phenomena based upon this conception involves a standpoint in which: (1) relations are stressed over parts; (2) the object of inquiry is the underlying set of transformational properties through which empirical phenomena are generated; and (3) the conscious human subject is viewed not as given, but as constituted by his or her subjection to pre-existent linguistic, mental, or social structures.

Notable scholars who, willingly or not, have been labeled as "structuralist" include Claude Levi-Strauss in anthropology, Jacques Lacan in psychoanalysis, and Michel Foucault in the history of discourse.[14] There have been, moreover, structuralist "readings" of

---

Golledge (New York: Methuen, 1981), p. 232.

11. Clifford Geertz, "From the Native's Point of View," in *Interpretive Social Science: A Reader,* ed. Paul Rabinow and William Sullivan (Berkeley: University of California Press, 1979), p. 229.

12. For useful overviews of structuralism, see Russell Keat and John Urry, *Social Theory as Science* (London: Routledge and Kegan Paul, 1975), chap. 6; Jean Piaget, *Structuralism* (New York: Basic Books, 1970); and John Sturrock, ed., *Structuralism and Since: From Levi-Strauss to Derrida* (Oxford: Oxford University Press, 1979).

13. Ferdinand de Saussure, *Course in General Linguistics* (London: Fontana, 1974).

14. Claude Levi-Strauss, *Structural Anthropology* (London: Allen Land, 1968); Jacques Lacan, *Écrits: A Selection* (New York: W. W. Norton and Co., 1977);

past intellectual projects undertaken long before the contemporary structuralist "movement" came into being. Most significant for the present discussion is the structuralist-influenced interpretation of Marx, which in recent years has become increasingly popular.[15] By demonstrating that many phenomena within capitalist economies are generated by unseen mechanisms associated with the mode of production, Marx in his mature work forged an approach to social analysis in which the human individual is no longer the central object of investigation. The remainder of this essay assesses the strengths and weaknesses of "structural-Marxism" as a basis for behavioral research in geography.

Marxism as Structuralism

In such early writings as his *1844 Economic and Philosophic Manuscripts,* Marx focused upon the ways in which capitalist wage-labor alienates and damages the worker.[16] These essays were "humanist" in their orientation, inasmuch as Marx's central concern within them was for "man," and the social barriers which prevent man from fully expressing his essence, or species-nature. Soon after the completion of these manuscripts, however, Marx's approach to the analysis of capitalism shifted. In the sixth of his "Theses on Feuerbach," he wrote: "The human essence is no abstraction inherent in each single individual. In its reality it is the ensemble of the social relations."[17] And again in his 1857 *Grundrisse* notebooks: "Society does not consist of individuals, but expresses the sum of interrelations, the relations within which these individuals stand."[18]

What Marx wished to emphasize was that all meaningful statements regarding concrete "personhood" involve inescapably social predicates. After 1845, then, the proper object of knowledge for Marx was neither the discrete human being nor the generic "man," but was instead the social *relations* through which specific forms of individuality are shaped.

Michel Foucault, *The Order of Things: An Archaeology of the Human Sciences* (New York: Pantheon Books, 1971).

15. Louis Althusser and Etienne Balibar, *Reading Capital* (London: New Left Books, 1970).

16. Karl Marx, *Economic and Philosophic Manuscripts of 1844* (Moscow: Progress Publishers, 1974).

17. Karl Marx, "Theses on Feuerbach," in *The German Ideology,* ed. C. J. Arthur (New York: International Publishers, 1970), p. 122.

18. Karl Marx, *Grundrisse,* vol. 1 (New York: Vintage Books, 1973), p. 265

As is well known, Marx's principal focus was upon relations of production, not only because people must eat before they can "make history," but because he felt that labor is the most essential, most characteristically *human* activity. It is through the creative appropriation and transformation of nature within the labor process that human beings transform their own nature.[19] Moreover, class relations crystallize through conflict over access to the means of subsistence and over disposition of the social surplus product. In sum, Marx believed that it is through the collective development of human productive powers within antagonistic class relations that history unfolds.[20] As an aspect of this evolutionary social development, new forms of human subjectivity continually come into being.

Although it is dubious whether these broad conceptions can be applied to all societies, Marx clearly did develop an indispensable framework for the study of social formations in which capitalism is the dominant mode of production. Marx viewed capitalism not as monolithically evil but as contradictory in character. Whereas it is a dynamic system through which humanity's productive capacities are increased to unprecedented levels, such powers expand in an uncontrolled fashion, according to the blind dictates of "the market." Furthermore, the creation of unprecedented wealth is predicted upon social relations which are exploitive, alienating, and even mystifying.

Marx determined that the essential character of capitalist production relations is actually quite different than it appears to everyday consciousness. On the surface, production proceeds on the basis of a freely-negotiated exchange of labor for wages, an agreement between independent and equal juridical subjects. The way that this wage-relation is "coded" and thus experienced, however, masks the underlying reality that workers continually reproduce their subservience to the capitalist through the creation of surplus value.[21] Workers are compelled neither by force nor by statute to toil for wages, yet for those who own no means of subsistence, there is really little alternative. For the proletariat, then, formal freedom masks not only exploitation but bondage.

---

19. Karl Marx, *Capital* (New York: International Publishers, 1967), p. 177.

20. For an excellent discussion of Marx's historical materialism, see G. A. Cohen, *Karl Marx's Theory of History: A Defence* (Princeton: Princeton University Press, 1978).

21. Cf. Karl Marx, "The Transformation of the Value (and Respectively the Price) of Labour-Power into Wages," *Capital,* vol. I, part VI, chapter XIX, pp. 535-542.

On the other hand, those whose livelihood is gained through appropriation of the surplus produced by workers presumably have the time and means to become more self-determining. Marx emphasized, however, that the coerciveness of the capitalist system extends even to the wealthy. Since the overall accumulation process is anarchical and prone to periodic crises, all capitalists live in constant danger of falling into the abyss of failure and poverty. Consequently, no matter what a particular factory owner may prefer, he or she is forced through competition to enforce labor discipline and to act otherwise as "the boss." Furthermore, investors feel incessant pressure to direct their capital into endeavors which will provide maximal returns, no matter where located, and no matter what the cost in terms of human well-being or environmental damage.

As Marx describes at length in his works on political economy, the transformation of value into its various forms, and the cyclic rhythms of accumulation and crisis all proceed in accordance with a systematic "logic" to which nominally free individuals can but constantly react. People are socialized into a pre-existent set of property and class relations which determines--in the sense of exerting pressures and setting limits--what they can or cannot do or become.[22] As a result, the decision-making processes which are of such fundamental concern to behavioral-perceptual geographers must be viewed as largely derivative, with both objective choice and subjective motivation shaped by the emergent properties of what Marx metaphorically terms "the social mechanism."[23]

By now, Marx's implied solution to our geographical "whodunit?" is more apparent. Perceiving, thinking individuals are compelled by social forces over which they have little control and of which they may even be unaware. Far more determined than determining, the constitutive human subject is effectively "de-centered" within social analysis. Instead, it is the generative properties, or "deep structures" of the capitalist mode of production and their systematic consequences which are the principal objects of inquiry. As the title of Marx's three-volume *magnum opus* suggests, it is "capital," and not conscious human agency, which propels history in societies like our own. This distinctive focus upon laws and tendencies originating within prevailing property relations provides the methodological foundation for a structuralist Marxism.

---

22. Cf. Raymond Williams, *Marxism and Literature* (New York: Oxford University Press, 1977), pp. 83-89.

23. Karl Marx, *Capital,* vol. I, p. 592.

Within this broadly "anti-humanist" viewpoint, there are at least two distinguishable variants from which geographers have drawn.[24] The work on urbanism by David Harvey, for example, has much in common with the "capital-logic" approach of Ernest Mandel and Roman Rosdolvsky, in which crucial importance is attributed both to the tension between use-value and exchange-value in the single commodity, and to the struggle between capital and labor in the production process.[25] Focusing narrowly upon Marx's critique of political economy in *Capital,* Harvey posits fairly direct connections between the dialectical "logic" of accumulation as presented by Marx, and the macro-scale evolution of the "built environment" within capitalist societies.[26]

Eschewing this view of society as an Hegelian "expressive totality" emanating from essential contradictions within the sphere of production, French philosopher Louis Althusser has pioneered a version of Marxism in which the non-economic aspects of social relations feature prominently.[27] Althusser argues that although economic structures ultimately are determinant in history, political and ideological structures display a "relative autonomy" that economistic Marxists fail to recognize. Both the nuances of Althusser's highly controversial reading of Marx and its application to the study of urbanism by Manuel Castells are beyond the scope of this essay.[28] Althusser's innovative treatment of ideology, however, is highly relevant to the issue of agency versus structure. For Althusser, ideology is not simply a set of false ideas regarding social reality, but is a "lived," constitutive element of social relations.[29] Ideology encompasses the experiential aspects of economic, civil, and familial relationships which in essence are objective, i.e., *not* defined in terms of consciousness and its

---

24. For an overview of the role of Marxist thought within American geography, see Richard J. Peet, "The Development of Radical Geography in the United States," *Progress in Human Geography* 1 (1977): 240-263.

25. Ernest Mandel, *Late Capitalism* (London: New Left Books, 1975); Roman Rosdolsky, *The Making of Marx's Capital* (London: Pluto Press, 1977).

26. David Harvey, *Social Justice and the City* (Baltimore: Johns Hopkins University Press, 1973), also "Labor, Capital, and the Class Struggle Around the Built Environment in Advanced Capitalist Societies," *Politics and Society* 6 (1976): 265-295; and "The Urban Process under Capitalism: A Framework for Analysis," *International Journal of Urban and Regional Research* 2 (1978): 101-132.

27. Althusser and Balibar, *Reading Capital*; also Louis Althusser, *For Marx,* (New York: Vintage Books, 1970); and Nicos Poulantzas, *Political Power and Social Classes* (London: New Left Books, 1968).

28. Cf. Manuel Castells, *The Urban Question* (Cambridge: MIT Press, 1977); for criticism and evaluation of Althusserian Marxism, see Norman Geras, "Althusser's Marxism: An Account and Assessment," *New Left Review* 71 (January-February 1972): 57-86; and Michael H. Best and William E. Connally, "Politics and Subjects: The Limits of Structural Determinism," *Socialist Review* 48 (November-December 1978): 75-99.

29. Louis Althusser, "Marxism and Humanism," in *For Marx,* p. 233.

properties. Althusser argues further that from birth human individuals are transformed into "subjects" through their ongoing practices within these relations. Hence, prevailing forms of consciousness, including the conception of one's self as free, whole, and autonomous, are themselves an "effect" of the ideological structure.

Particularly as developed by Althusser, structural-Marxism offers certain advantages over current behavioral-perceptual approaches. Because it focuses upon the essential generative properties of the capitalist mode of production, such a framework can account for emergent, macro-level phenomena without reducing them to the sum of individual actions. Moreover, structural-Marxists do not accept "consciousness" as an empirical given, but instead attempt to demonstrate how human subjectivity is constituted socially.

The Limits to Structural Marxism

Structural Marxism, however, is not without notable shortcomings. Because it is often pitched at a high level of abstraction, structural-Marxist analysis frequently presents both a reified, reductionist view of society and an impoverished view of the human individual. To explain why this is the case, it is necessary to delve more deeply into some of Marx's own formulations.

A crucial feature of Marx's critique of capitalism was his discussion of the fetishism of commodities and capital, and its converse, the reification of human subjectivity.[30] In an ideological inversion rooted in prevailing property arrangements, "capital" is treated discursively as active and productive, whereas "labor" is viewed as an object, another "factor of production." Overall, under capitalism, the emergent outcome of human activity seems to possess a "life of its own." As Marx described it in the *Grundrisse*:

> The social character of activity, as well as the social form of the product, and the share of individuals in production here appear as something alien and objective, confronting the individuals, not as their relation to one another, but as their subordination to relations which subsist independently of them and which arise out of collisions between mutually indifferent individuals. The general condition of activities and products, which has become a vital condition for each individual--their mutual interconnection--here appears as something alien to them, autonomous, as a thing.[31]

This passage suggests why Marx's own analyses were often presented in terms of the characteristics of emergent entities such as "capital" and "labor." For Marx, such entities are "autonomous,"

---

30. Cf. Georg Lukacs, "Reification and the Consciousness of the Proletariat," in *History and Class Consciousness: Studies in Marxist Dialectics* (Cambridge: MIT Press, 1971), pp. 83-222.

31. Karl Marx, *Grundrisse*, p. 157.

but only at the level of appearance. Such appearances, however, are *not* mere illusions, but are "real" inasmuch as under capitalism appearances are *lived,* and hence people are "ruled by abstractions."32

Within structural-Marxist anaysis, however, the human origins of emergent entities are not given adequate treatment. In most such writings, people are treated primarily as representatives of class positions, or as "personifications" of economic categories. As James Duncan and David Ley have observed, to the extent that any psychological processes at all can be inferred from the work of structural-Marxists in geography, what apparently is postulated is a simple "stimulus-response" model in which cognition itself is "black-boxed."33 Any plausible treatment of social life must account for the reflective, meaningful nature of human action. Certainly, meanings should never be treated as the *sine qua non* of social existence; but neither are they merely epiphenomenal. In other words, the structuralist "de-centering" of the constitutive subject should never, in the words of Anthony Giddens, "be made equivalent to its disappearance."34

Because structural-Marxists fail to incorporate a plausible framework for the explanation of human action into macro-level analysis, their quite proper emphasis upon reification as a social phenomenon is often obscured by a tendency toward reification within the mode of presentation itself. Indeed, as Duncan and Ley have observed, within structuralist-Marxist discourse "the mode of production makes demands, capitalism devises solutions, capital throws its weight, [and] social formations and modes of production write history."35 However, despite the need to recognize the causal efficacy of emergent entities, our geographical "whodunit?" can never properly be recast as a "whatdunit?"

The difficulties posed by this reification are compounded by a propensity to portray the "logic of capital" not merely as a central focus within a broader analytical framework, but as virtually synonymous with the whole of social existence. English historian E. P. Thompson has complained that for too many Marxists the vocabulary

---

32. Carol C. Gould, *Marx's Social Ontology* (Cambridge: MIT Press, 1978), p. 39.

33. James Duncan and David Ley, "Structural Marxism and Human Geography: A Critical Assessment," *Annals of the Association of American Geographers* 72 (March 1982): 40.

34. Anthony Giddens, *Central Problems in Social Theory* (Berkeley: University of California Press, 1979), p. 45.

35. James Duncan and David Ley, loc. cit., p. 36.

of nineteenth century Marxian analysis has become an eternal set of frozen categories mapped indiscriminately onto an ever-changing reality, with the mode of production itself becoming

> like a base camp in the arctic of theory which the explorers may not depart from for more than a hundred yards for fear of being lost in an ideological blizzard.36

Although Althusser avoids the monism of the capital-logic approach by postulating three "relatively autonomous" structures, he does not fundamentally avert the pitfall of reductionism. In the absence of more thoroughgoing attempts to explicate the delicate linkages between theoretical abstraction and the material reality to which it corresponds, structural-Marxists remain guilty on the whole of what Alfred North Whitehead termed "the fallacy of misplaced concreteness."37

Toward a Structurationist Alternative

From these telling criticisms, it might be concluded that this "anti-humanist" version of Marxism has been fundamentally misconceived. Consequently, an alternative might be sought within the distinguished parallel tradition of humanist Marxism. Indeed, a veritable humanistic broadside against structuralist approaches was delivered in 1978 by E. P. Thompson. In *The Poverty of Theory*, Thompson denounced Althusser at length, proclaiming against all attempts at structural explanation that "men and women are the ever-baffled, ever-resurgent agents of an unmastered history."38 For Thompson, the central analytical concept within historical materialism is social class, and the fundamental process for investigation is class struggle. Although he admits in his well-known book, *The Making of the English Working Class,* that the class experience is largely determined by the productive relations into which men are born--or enter involuntarily," Thompson contends that "class-consciousness," or "the way in which these experiences are handled in cultural terms," is not.39 Hence, classes "make" themselves through people's mutual recognition of common interests and modes of life, as much as they are "made" by objective circumstances.

---

36. E. P. Thompson, *The Poverty of Theory and Other Essays* (London: Merlin, 1978), p. 346.

37. According to Whitehead, this fallacy involves "neglecting the degree of abstraction involved when an actual entity is considered merely so far as it exemplifies certain categories of thought," *Process and Reality* (London: MacMillan, 1929), p. 11.

38. E. P. Thompson, *The Poverty of Theory,* p. 280.

39. E. P. Thompson, *The Making of the English Working Class* (New York: Vintage Books, 1966), pp. 9-10.

Because Thompson emphasizes that "social being" conditions but does not determine "social consciousness," some geographers claim that his work can contribute more to geographical research than can structural Marxism.[40] In a rejoinder to Thompson, however, Perry Anderson argues persuasively that the English historian attributes excessive autonomy to "consciousness." Marx himself stressed repeatedly that class relations are constituted not by thoughts and feelings but by the objective places and functions occupied by human beings within a mode of production. It would be reductionistic to claim that such relations mold consciousness in a direct manner, but to minimize as Thompson does, their determining effects entails an indefensible voluntarism. Overall, according to Anderson, Thompson exaggerates the role of agency in human history, "in defiance of the millenial negations of self-determination in the kingdom of necessity."[41]

It is a testament to the richness, if not the ambiguity of Marx's thought, that debate will not likely end soon regarding his dictum that "Men make their own history, but not of their own free will."[42] But although the force of Thompson's critique of structural-Marxism must be acknowledged, his uncritical acceptance of "experience" as, in Anderson's words, "the privileged medium in which consciousness of reality awakens and the creative response to it stirs" weakens the materialist, relational thrust of Marx's own work.[43] Hence, although revision of existing structural-Marxist frameworks is surely in order, the decentering of the conscious subject initiated by Marx must be preserved.

If a more satisfactory basis for structural-Marxist research is to be established, it must first be made clear that the so-called "logic of capital" represents a partial abstraction from any historically-specific capitalist society. That Marx in fact intended for *Capital* to serve as a starting point rather than a strait-jacket for empirical research is suggested by the complexity of his own historical writings, as well as by his frequently cited remark that "the concrete is concrete because it is the concentration of *many* determinations, hence unity of the diverse."[44] Clearly, what is implied is that any historical "conjuncture" (i.e.,

---

40. Cf. James Duncan and David Ley, "Structural Marxism," p. 54.

41. Perry Anderson, *Arguments Within English Marxist* (London: New Left Books Review and Verso Editions, 1980), p. 58.

42. Karl Marx, "The Eighteenth Brumaire of Louis Bonaparte," in *Surveys from Exile,* ed. David Fernbach (New York: Random House, 1973), p. 146.

43. Perry Anderson, op. cit., p. 57.

44. Karl Marx, *Grundrisse,* p. 101.

set of events) involves determining factors other than the economic.[45] Not only must the political and ideological "instances" posited by Althusser be delineated, but so must the psychological processes that are of interest to behavior-perceptual geographers. Although specific forms of subjectivity are indeed conditioned socially, the result of this socialization is that people do possess "causal powers" to invoke the terminology of the realist theory of science.[46]

To admit that social and spatial outcomes involve causes that are non-economic does not mean, as Duncan and Ley imply, that non-reductionist Marxian research becomes virtually indistinguishable from Weberian approaches.[47] As Roy Bhaskar argues, an "integrative pluralism" can incorporate numerous levels of causation while maintaining, in contrast to Weber, a firm grounding within the "material bases of human life."[48]

A second revision that must be incorporated into future structural-Marxist research pertains to the troublesome concept of "structure" itself. A consistent problem in dealing with this term is that it has tended to connote an external, absolute constraint upon human behavior. Such contemporary social theorists as Pierre Bourdieu, Alain Touraine, and Anthony Giddens, however, emphasize that "structure" is enabling as well as constraining.[49] Within Giddens' "structuration" approach, for example, structure is viewed "not as a barrier to action, but as essentially involved in its production."[50] Giddens stresses that structure is not a thing-like, reified entity with an independent ontological status, but consists of a generative set of rules and resources from which individuals draw during the course of their ongoing social interaction. Thus, structure "exists" only by virtue of its actual reproduction and transformation within simulated human practices. This Giddens refers to as the "duality of structure"--the fact that structure is

---

45. Cf. Andrew Sayer, "Abstraction: A Realist Interpretation," *Radical Philosophy* no. 28 (Summer 1981), pp. 6-15.

46. R. Harre and E. H. Madden, *Causal Powers: A Theory of Natural Necessity* (Totowa, N.J.: Rowman and Littlefield, 1975); also Roy Bhaskar, *The Possibility of Naturalism* (Atlantic Highlands: Humanities Press, 1979).

47. James Duncan and David Ley, "Structural Marxism," pp. 53-54.

48. Roy Bhaskar, "Emergence, Explanation, and Emancipation," in *Explaining Human Behavior: Consciousness, Human Action and Social Structure,* ed. Paul F. Secord (Beverly Hills: Sage Publications, 1982), p. 278.

49. Pierre Bourdieu, *Outline of a Theory of Practice* (Cambridge: Cambridge University Press, 1977); Alain Touraine, *The Self-Production of Society* (Chicago: University of Chicago Press, 1977); Anthony Giddens, "Notes on the Theory of Structuration," in *Studies in Social and Political Theory* (New York: Basic Books, 1977), pp. 129-134.

50. Anthony Giddens, *Central Problems in Social Theory,* p. 70.

at one and the same time the "medium and the outcome of the social practice it recursively organizes."[51]

Within the structuration perspective, people are not merely supports for structurally-defined roles, but are capable actors whose meaning-endowing, reflective consciousness and stocks of knowledge play an irreducible role within the reproduction of social systems. This "knowledge," however, is not always discursively available; it is in large part tacit, practical knowledge of social conventions and modes of conduct which cannot be verbalized.

At first appraisal, structuration may appear to resolve the tension between agency and structure in favor of the former.[52] In Giddens' view, however, social life remains highly coercive. First, although people act in accordance with a sort of "rule-governed creativity," they are, in fact, unaware of many of the rules. Second, there is clear acknowledgement that resources, and therefore power, are unequally distributed in society. Finally, since society is a structured, interdependent totality, each individual human action occurs within a context of unacknowledged conditions and produces unintended consequences. Hence, Marx's insights regarding an emergent, systemic logic irreducible to the volitional activities of individuals remain both valid and necessary. In short, whereas people may not be structural puppets, neither are they masters of their own fate.

Prospects of Behavioral Research

For research applications, a great strength of the structuration approach is that it encourages Marxian scholars to complement abstract, macro-level analysis with concrete investigations into the everyday practices which comprise the fine texture of human history. It must be demonstrated how the structural properties of a social system are reproduced or transformed within specific, territorially-bounded contexts. Geographers can examine, for example, how a global capitalist system evolves through myriad local alterations in labor processes and employment patterns, or through flows of capital into and out of property markets. These phenomena, in turn, can be linked with the ongoing transformation of landscapes.

---

51. Anthony Giddens, "Agency, Institution and Time-Space Analysis," in *Advances in Social Theory and Methodology,* ed. K. Knorr-Cetina and A. V. Cicourel, p. 171.

52. This is claimed by Derek Layder, *Structure, Interaction and Social Theory* (London: Routledge and Kegan Paul, 1981); also Nigel Thrift, "On the Determination of Social Action in Space and Time," *Society and Space* 1 (forthcoming, 1983).

Such research can draw from a variety of information sources, including government documents, local newspapers, personal diaries, and informal interviews. The data thus generated, however, never simply "speak for themselves," but must be treated as a sort of raw material within a broader process of knowledge-production.[53] Since "facts" frequently are theoretical constructs, the internal relationships linking researcher, concepts, and the material world must constantly be explored. Such attentiveness to epistemological and ontological issues is particularly crucial for practitioners of a "realist" structuralism in which the aim is the discovery of unseen causal mechanisms which generate overt phenomena. If such research is to avoid lapsing into sheer speculation, the rigorous dialogue between fact and theory must be maintained vigilantly.

For the moment, specific examples of structuration-influenced research remain scarce. However, in a series of recent essays, Nigel Thrift and Allan Pred have explored the possible compatibility of structuration with time-geography.[54] In a study conceived in part with this synthesis in mind, Susan Christopherson has shown how a global transformation in the division of labor within the electronics industry has influenced the lives and consciousness of young working-class women in Juarez, Mexico.[55] Accounting carefully for both the social-structural determinants and the time-space bound daily and life paths through which individual experiences are shaped, Christopherson demonstrates how individual and social change are dialectically interconnected within a specific locale.

At a larger territorial scale, Derek Gregory has called for the establishment of a "reconstituted regional geography" based upon a structuration perspective. Gregory believes that a regional approach can best demonstrate how "spatial structures are implicated within social structures, and each has to be theorized with the other."[56] He has provided an example of such research by investigating the development of the woollen textile industry in Yorkshire during the Industrial Revolution.[57]

---

53. Cf. Althusser and Balibar, *Reading Capital*; for a contrasting view, see Peter Gould, "Letting the Data Speak for Themselves," *Annals of the Association of American Geographers* 71 (June 1981): 166-176.

54. Nigel Thrift and Allan Pred, "Time-Geography: A New Beginning," *Progress in Human Geography* 5 (1981): 277-286; Allan Pred, "Structuration and Place: On the Becoming of Sense of Place and Structure of Feeling," *Geografiska Annaler* 63B (1981): 5-22.

55. Susan Christopherson, "Family and Class in a New Industrial City," unpublished doctoral dissertation, Department of Geography, University of California, Berkeley, 1982.

56. Derek Gregory, *Ideology, Science and Human Geography* (London: Hutchinson and Co., 1978), p. 172.

57. Derek Gregory, *Regional Transformation and Industrial Revolution*: A

Structuration can, moreover, inform studies of land-use conflicts and other aspects of what might be termed local historical geography. The author's own doctoral research, for example, has examined two suburban California communities where rapid, uncoordinated development during the 1960s resulted in environmental and fiscal problems which precipitated the establishment of growth controls.[58] To explicate these complex events, it was necessary to determine and disaggregate their many distinct causes. In brief, the diffuse structural properties of American capitalism coalesced with more contingent local institutional factors to create conditions in which class-based political conflict over "growth" became possible.

This controversy, however, could not be explained without reference to people's values and perceptions regarding their local communities. Consequently, a focus upon structure was complemented by a concern for ideology. First, it was discovered that development during the 1960s was facilitated by a pro-growth boosterism promulgated by local businessmen, realtors, and newspaper publishers. Members of this "growth coalition" were able to use their economic power to successfully represent their own interests as those of the community as a whole. Second, it was determined that when opposition to growth did emerge among local intellectuals, the dissenters still believed "the system" to be essentially just. In part, this is because, as Marx showed, the surface appearances through which economic life is experienced systematically mask its exploitive aspects. Since fundamental questions regarding private property as an institution were rarely raised, political activism in opposition to growth was limited to a mild liberal reformism.

Although all of these examples focus upon specific locales, they still represent a "bird's-eye" perspective when contrasted with most contemporary behavior-perceptual geography. It remains to be seen to what extent a Marxian approach can account adequately at the individual level for the internal mental processes which mediate between environment and behavior. The possibility that there can be a specifically Marxian psychology has been explored in the work on personality by Lucien Seve, and in the cognitive research of A. R. Luria and L. S. Vygotsky.[59] Within such a socially-informed

---

Geography of the Yorkshire Woollen Industry, 1780-1840 (London: MacMillan, 1980).

58. Douglas Greenberg, "Growth and Conflict at the Suburban Fringe: The Case of the Livermore-Amador Valley," unpublished doctoral dissertation, Department of Geography, University of California, Berkeley, 1982.

59. Lucien Seve, *Man in Marxist Theory and the Psychology of Personality* (Atlantic Highlands: Humanities Press, 1978); A. R. Luria, *Cognitive Development: Its Cultural and Social Foundations* (Cambridge: Harvard University Press, 1976);

approach to the human individual, in Luria's words,

> The basic categories of human mental life can be understood as products of social history--they are subject to change when the basic forms of social practice are altered and thus are social in nature.60

In other words, consciousness is always contextual, rooted within material relations involving class and power. As Marx and Engels wrote in *The German Ideology,* "consciousness can never be anything else than conscious existence, and the existence of men in their actual life-process."61

In conclusion, I have suggested here that an essentially "structural" Marxism need not eliminate totally the conscious human subject. However, both the programmatic nature of this essay and the indisputable complexity of the world in which we live dictate that no final solution can be offered to the "whodunit?" in which we are involved as human geographers. That the effectivity of human agency is so clearly in doubt, however, only underscores the need for continuing detective work that not only is scholarly and careful, but is also committed to revealing and overcoming the many social constraints that now fetter free and creative human subjectivity.

L. S. Vygotsky, *Thought and Language* (Cambridge: MIT Press, 1962).

60. A. R. Luria, *Cognitive Development,* p. 164.

61. Karl Marx and Fred Engels, *The German Ideology,* ed. C. J. Arthur (New York: International Publishers, 1970), p. 47.

CHAPTER 14
SEEING, THINKING, AND DESCRIBING LANDSCAPES

Edward Relph

At the outset, let me make my perspective on behavioral geography clear. Since I have never been able to establish just what 'behavioral geography' is and how it distinguishes itself from other sorts of geography, I have assumed it to be a version of B. F. Skinner's behaviorism somehow transferred from psychology to geography. Behaviorism has to do with the conditioning of actions by the application of appropriate stimuli--electric shocks, mirrors in shopping malls, muzak in offices, and so on. These stimuli can be used to generate desired patterns of behavior and thus to promote productivity and profits. Moreover, in so far as individuals are unaware that there are such stimuli and that they are responding to them, this behavioral manipulation denies the free will of everyone except the manipulators. Because of this, I am deeply suspicious of behavioral approaches, whether in geography or other disciplines, though I acknowledge that they may have a limited value in psychiatric treatment. One of my main concerns in teaching is in fact to alert my students to the possibility that their behavior may be being manipulated by various insidious means such as variations in light intensity to change mood, plastic-molded chairs comfortable for no more than ten minutes in fast-food restaurants, or propaganda campaigns to create a pleasant sense of place and community.

It may be protested that behavioral geography has no such manipulative intentions and seeks only to observe and explain the patterns of spatial and environmental behavior. However, my suspicions run so deep that I regard any attempt at explaining human behavior as the first step towards the control of that behavior, just as explanation of natural processes lead inexorably to interventions in those processes. Grasp the structure of the atom and split the atom, identify DNA and clone plants. The difference between natural science and social science in this respect is that whereas explanation in natural science is transferred to technologies for the human domination of nature, social science assists the control by political means of some groups of society by those other groups which are privy to the explanations. This is putting it dispassionately; Noam Chomsky was far more bitter when he wrote that "as the scientists are busy engineering the world's annihilation, the social scientists have been entrusted with the smaller mission of engineering the world's consent."[1]

There are no innocent parties of any of this. If geographers in their role as consultants choose *not* to apply their own explanations of spatial and environmental behavior, then someone else will surely borrow these and turn them to whatever ends they can. In his book, *The Arrogance of Humanism,* David Ehrenfeld proposes a law of experience which states that whenever a scientific discovery can be used for evil or mischievous purposes it certainly will be.[2] The consequences of behavioral geography may be less horrendous than those of atomic physics or even psychology, but the results can still contribute to the loss of human freedoms by assisting, for instance, in the design of communities in which the efficiency of spatial behavior is maximized. I occasionally have grey visions of a country called Isotropia, a uniform plane of uniform central places, with uniform people living and working in uniform monotony to maintain perfect efficiency of location and production. In short, in so far as behavioral geography is concerned with explaining human behavior, it is ethical; and in so far as those explanations can be used in reconstructing places or communities, it is political. Arguments to the effect that it is a detached and disinterested branch of social science warrant only derision.

These criticisms may be further protested on the grounds that what I am describing is not "behavioral" so much as "behavioristic," that behavioral geography is concerned with ethical and political consequences and that it is only the crass extensions of behavioral approaches by unthinking hacks and scribblers who are to be condemned. There is merit in this protest in that it draws a neat theoretical distinction between those who approach their research with a full awareness of its possible implications and abuses, and those who proceed with a naive faith in the neutrality and glory of scientific method. However, in practice, I do not know how such a distinction could be pressed. No one is going to admit to being a behavioristic geographer, so it would simply be a holier-than-thou term of abuse signifying nothing. I find that I can manage

---

[1]. Noam Chomsky, *American Power and the New Mandarins* (New York: Pantheon Books, 1967), p. 110.

[2]. David Ehrenfeld, *The Arrogance of Humanism* (New York: Oxford University Press, 1978), p. 97.

perfectly well in my own work without reference to either behavioral or behavioristic geography, and I intend to continue thus managing.

So I do not consider myself to be a behavioral geographer, though others may choose to categorize me in this way. Since they presumably understand something positive and precise by this description, and I frankly confess my ignorance as to its meaning, it is entirely possible that they are correct. However, to my considerable inconvenience in explaining what it is that I do to others, I find myself obliged to eschew all such labels because they are misleading and constraining. I do not consider myself to be a "humanistic geographer" or a "phenomenological geographer," or any of the other more traditional types of geographer. The very idea of such slots makes me uneasy. Indeed, for a long time I quietly disowned the name "geographer," albeit on the entirely false grounds that it suggests someone intellectually lackluster and shallow in comparison with, for example, economists and physicists. Then I realized that economists and physicists are respectively engaged in trying to destroy our way of life and our entire planet with their hard-headed theories and atom splitting. I now consider it to be rather pleasant to be a geographer studying a subject incapable of exact definition, contributing little to this process of desolation, and possibly even beginning to disclose gentle, non-manipulative ways of delaying it.

The specific directive I have for writing this paper is to say something about the approaches I use in my research. At the risk of pontificating, I have taken this directive literally and have tried to reflect on my approach to geography and to be as straightforward and as honest as I can in describing it. So, this is not a formal methodological account in the manner of yet another outline of scientific or phenomenological method--that is, logical, abstract, compendiously referenced and quite impracticable. Nor is it one of those essays that proselytizes a particular philosophical perspective. Such outlines and essays invariably leave me and my students utterly bemused. I have a curious inability to formulate hypotheses, construct tests, suspend presuppositions or to bracket off ideas, as these tasks are required by various methodological dictates. And I am unmoved by appeals for the necessity of adopting specific philosophical postures like idealism, pragmatism, Marxism, or, heaven forfend, positivism. Indeed, I often find that the substantive writing of those who advocate such postures produces results that could equally well have been achieved by the application of common sense without bothering with all the philosophizing.

Methodology and philosophy are deeply implicated in my research and are not isolated in some rarified realm of discourse. I did not choose my philosophical perspectives through a rational decision-making procedure which compared the merits of Marxism, positivism, phenomenology, critical theory and so on, then identified the best of these. Nor, I trust, did anyone else. In fact the opposite is almost true--the philosophies and methods I use seemed to choose me. Consequently, what I write here can perhaps be best described as a fragment of an academic autobiography in which I attempt to reflect on and reveal the philosophical and other cornerstones of my approach to geography.

Seeing, Thinking and Describing

I believe the major insights of geography are these: that there are no clear boundaries to natural or social processes; that nothing happens with absolute certainty or in isolation; that ordinary places and landscapes are far more complex than anything encountered in an econometric model or a physicist's laboratory. Given these insights, I have to recognize that this "bloomin', buzzin' confusion," as William James once described it, can never be neatly explained nor totally controlled. In this there is hope. Meanwhile, ongoing efforts at such explanation lead to crass simplification, and attempts at control result in nasty unanticipated effects and deprivations. For my part, I try to keep my sights low and to focus on what is particular and close at hand even as I reflect on the remote and subtle consequences of events and their philosophical and ideological contexts. An appropriate slogan for this is the call by Rene Dubos to "think globally and act locally."[3] For me, this requires looking carefully at landscapes, allowing my thinking to respond to what I see, and then attempting to describe clearly the character of these landscapes.

These three procedures of seeing, thinking and describing are analytic categories; that is to say, they are distinguishable by reflection rather than being parts of a deliberate practice. They are not separable activities, but overlap and come in no special order. The activity of describing a place may cause me to think about it differently; my reflections on a landscape may lead me to come to see it in a new light; and it is impossible to see without thinking, or to describe without seeing at least in one's mind's eye. However, my approach can be presented more lucidly if I treat seeing, thinking and describing separately.

---

3. Rene Dubos, *The Wooing of Earth* (New York: Charles Scribner's Sons, 1980), p. 156.

Seeing has to do with direct observation and ensures that I write about things that I have studied for myself rather than gleaned from other people's books or statistics. By thinking, I mean the attempt to make sense of what I have seen on its own terms and not by reference to *a priori* hypotheses or borrowed models. Description is simply the process of expressing what I have seen, of putting together observations and reflections into a sensible statement. The academic philosophizing and debates about methodology not uncommon in geography have little value for seeing, thinking and describing as I understand them. Such discussions are too detached, too concerned with points of definition and fine distinctions between categories; they are caught up in abstract analysis and appeals to authority. This is not to say that philosophical speculation is unimportant, and in fact I am convinced that work lacking a sound philosophical foundation is bound to be shallow and ephemeral. But philosophy is more than an activity reserved for academic research to be left behind in the university library at the end of the day; it is rooted in and related to all aspects of life. I acknowledge that this may not be so obvious or possible in all fields, including, for example, geomorphology, but for me there has to be some continuity between what I profess in my teaching and writing and the way I live my life as a citizen, father and property owner.

The particular environments which interest me are landscapes--especially modern, man-made ones, for I live in the midst of these. Landscapes I understand to include almost all the features of the world encountered visually: trucks, petunias, clouds, airports, streets, garbage cans, automobiles, people. Landscapes have an immediate and sensible, if somewhat elusive, reality about them, so I have something definite to see, think about and describe. In this small respect, my work corresponds with that of a physical scientist because I am dealing with some *thing* rather than with ideas, images and institutions. However, unlike a scientist who analyzes matter into its constituent parts and processes, I have to try to grasp and convey the entire shifting scene. To divide it into building, sidewalks, roadways, parking meters and so on, is to lose the wholeness that is the essence of landscape. This holistic quality is significant because it renders landscape incapable of any sort of quantitative analysis.

It is impossible to ignore landscape, for it frames our lives. It is the totality of what we encounter out-of-doors (and through windows and in shopping malls, sports domes and other vast enclosed spaces). It can be depicted in drawings, paintings and photographs; it can be shown to others; the qualities of specific landscapes can be identified and debated. Yet it is possible neither to delimit a landscape, not to walk around it rather than through it, nor to say where one landscape ends and another begins. Given this indeterminacy it should be clear why I confine my efforts to description.

In describing landscapes, I am not concerned with cognitions or perceptions other than my own. The only images which interest me are those manifest in symbols like those of democracy in state capitol architecture, or the images of aspiration and nostalgia that can be seen in dwarf-jockeys, wagon-wheels and trim lawns in suburban gardens. A conventional approach adopted in behavioral geography is to survey the perceptions, images and attitudes of others. My concern, however, is to communicate what *I* see and think about a place. I try not to generalize about what others may see or think, though I do attend to what has been written about the place or similar places, especially if it is based on direct observation.

This approach may seem to be unduly subjective, idiosyncratic and egocentric--a vain way of doing things that will product only impressionistic results. It is subjective and idiosyncratic if those terms mean that I must stand or fall by the quality of my thinking and my descriptions. I have no impersonal, borrowed set of methods which I can use to absolve myself of responsibility for the inadequacy of my ideas or as an excuse for thinking. It is, however, not egocentric if what I describe makes sense and clarifies matters for others, and it is not impressionistic if the seeing and thinking I practice can penetrate the surface forms of landscapes to reveal something of their inner character or *genius loci*. I am not trying to flaunt my petty foibles and prejudices, just to write about something I have studied so that it makes sense to others.

Seeing

Seeing or looking carefully at landscapes is not a simple matter. Of course, all sighted people see in the trivial sense of noticing doorways, traffic signals and the like, but at its other extreme, "seeing" means insight and understanding. A "seer", for example, is someone with visionary powers. What I am concerned with is seeing as a faculty that can be made increasingly more sensitive in order to attain improved understanding. By a deliberate effort of concentration and wonder about the nature of things, it is possible to see in greater detail, to discern hues and textures, literally to develop a greater acuity of vision. Whether innately or through practice, many artists have highly developed skills of observation. The painter Whistler used to show off by looking for two minutes at a view he had come upon suddenly, around a street corner perhaps, then turning his back to it and recounting the whole scene in minute detail to his companions. Some art teachers maintain that it is impossible to draw well unless first you can see well, and one at least, Nikolaides in *The Natural Way to Draw,* has devised an entire series of drawing exercises intended to teach anyone who is sufficiently committed both to see and to draw well.[4] Most people never bother with such matters, of course, and the scruffiness and lack of finish in many new man-made landscapes suggests that most of us wander along with our eyes to the ground noticing only road signs, doorways and those scenes to which our attention has been called by guidebooks.

Since seeing is a difficult skill to learn, I have turned to John Ruskin as my teacher. Ruskin was a nineteenth-century art critic and social reformer who came to prominence for his defense of the painting of Turner and his condemnations of most of the achievements of the industrial age.[5] Turner's paintings of ships in storms, steam engines in rain, the burning of the House of Commons, were described by other art critics as messes presumably achieved by throwing breakfast at a canvas and therefore utterly unrealistic. Ruskin argued that, on the contrary, these paintings were based on painstakingly accurate observation. Unlike other landscape painters, notably Claude Lorrain, Turner painted the particularity of what he saw. Lorrain's canvasses show vistas framed by great trees, usually with some event from classical mythology acted out in the foreground. His trees are a kind of general tree and his rocks, general rocks. Lorrain's figures leave much to be desired. In one painting, Ruskin discovered an archer with an arrow so drawn back on his bow that the string when released would fall slack and with the arrow unsupported below the archer's bow hand. In contrast, Turner painted elms, oaks, beeches; he noticed and depicted the differences between granite and dolomite and schists; to make his observations for storms at sea, he on one occasion strapped himself to the mast during a gale. It was the critics who were wrong, Ruskin argued, because they were trapped in anachronistic visual conventions and

---

4. Kimon Nikolaides, *The Natural Way to Draw* (Boston: Houghton Mifflin, 1969).

5. John Ruskin's *Modern Painters,* published in five volumes between 1843 and 1860, and in many subsequent editions, deals with Turner and the idea of landscape. Volume III is especially valuable for its discussions of landscapes.

easy habits of seeing that led them to admire the sloppy and inaccurate generalizations of Lorrain's paintings. The message was, and still is, clear: thorough and accurate seeing requires a continuous expenditure of effort.

Throughout his writings on art and architecture, Ruskin stressed the necessity of seeing clearly, and he invariably related this to landscape, which he described as one of the enduring interests of his life. To see clearly, he suggested, it is necessary to develop an unpretentious attitude of vision that strives to omit nothing yet does not impose sentiments on what is seen. This vision does not endow nature with human and invested qualities, nor does it reduce landscapes to mere inventories of facts. Instead, it demands careful attention and open-mindedness to what exists in all its aspects and qualities. It was, he suggested, not always necessary to notice the qualities of individual objects, for that would become impossibly time consuming and miniscule; but the qualities and characteristics of types of objects should be noted, hence his praise for Turner's depiction of elms and oaks. Such careful observation requires that landscapes be approached slowly. Ruskin deplored trains, which he claimed turned people into parcels; he would have found cars and airplanes unimaginably loathsome. And finally, but certainly not least, clear seeing demands that we regard not just the surface forms of things but exercise our imagination to grasp connections and implications. A row of Weaver's cottages might be picturesque and pleasant to look at, but the lives of the families who occupied them were filled with hardship and poverty. That fact must be part of one's vision. Ruskin was involved in the origins of British socialist movement, and he was well aware of the dangers of relying on superficial appearances and of the need to take into account social and political matters.

Ruskin does not provide a set of rules or guidelines for seeing that can be learned and applied. The value of his writing is that it suggests the complexities and limitations of vision; it challenges my habits of seeing and causes me to consider other possibilities. Of course, he did write over a century ago and much of what he said must be understood in terms of a social context quite different from the present. Much of his advice depends on a slow, deliberately sustained act of visual perception. Many of the landscapes I encounter are seen through windows of rapidly moving vehicles, or at those brief moments of passivity while waiting for traffic signals to change. Though I do practice Ruskin's slow, imaginative approach whenever I can, much of my seeing is less incisive than this. It is a matter of always looking, always attending to the character of the place where I happen to be. I made no claim that this is in any way exceptional or unusual--it is perhaps one of the traditional methods of geographers, though it seems to have slipped into general disuse. It is an activity that can be called "watching"--sometimes for the typical, sometimes for the unusual. It can be done carefully and with direction, or it can be an adjunct to other activities. It is not mysterious or obscure or technical, but it is a skill that can be improved. Anyone who practices it seriously is unlikely to be content with first and superficial impressions: "watching" inexorably leads away from the tourists' paths and the monuments of civic pride; it leads into the suburbs of cities, into the back alleys and the grey concrete spaces under elevated expressways.

Thinking

Seeing involves what Ruskin described as a "curiously balanced condition of the powers of the mind."6 To see clearly, it is necessary to think clearly. Too much looking leads to a confused welter of impressions; too much thinking results in cold abstractions and false images being imposed onto places. Seeing and thinking must therefore be held in tension.

As Ruskin serves as my teacher for seeing, so Martin Heidegger, through his works of philosophy, serves as my teacher for thinking. These two may seem like strange bedfellows, since they are of different centuries, different countries and different intellectual backgrounds, yet in their writings I find many similarities. It seems that almost every idea Ruskin expressed about seeing is reinforced and given depth by Heidegger's explorations of thinking.7 While Heidegger's works cover many issues and take many turns, a central theme has to do with what constitutes an appropriate way of thinking. The answer he discloses is that appropriate thinking responds to the essential character of whatever is being thought about. It imposes no fixed methods, logics or stratagems but allows things to manifest themselves in their own being. This does not mean that one's mind has to be considered a *tabula rasa,* a blank sheet absorbing whatever the world happens to

---

6. Ruskin, *Modern Painters,* vol. 3, chapter 17, section 5.

7. Heidegger's classic work is *Being and Time* (New York: Harper and Row, 1962); section II.7 in the introduction is about "The Phenomenological Method of Investigation," and I have found this particularly useful. Especially valuable secondary sources on Heidegger are W. J. Richardson, *Heidegger: Through Phenomenology to Thought* (The Hague: Martinus Nijhoff, 1967); and Walter Biemel, *Heidegger: An Illustrated Study* (New York: Harcourt, Brace, Jovanovich, 1973).

present to it; we all have biases, recollections and intentions which condition our thinking. What we have to do is to make ourselves aware of these and then try to understand how they influence our ways of thinking. Furthermore, nothing is assumed away in advance as being unreasonable or irrelevant, simply because it is impossible to know in advance just what is unreasonable or irrelevant.

In one of his lectures, Heidegger called this "meditative" or reflective thinking, and contrasted it with "calculative" thinking.[8] Calculative thinking analyzes, computes, organizes, classifies, and measures; it does not pause to reflect upon itself as it busily restructures the world according to its well established procedures of instrumental reason. It probes the structure of matter, finds cures for diseases, explains the patterns of human behavior; then it shatters atoms, synthesizes chemicals of horrible toxicity, and strives to control human activities. Meditative thinking, on the other hand, can offer no systems and no quick remedies. It is, however, not an autistic retreat from reality. In fact, it demands a long and careful effort of self-consciousness about one's own thought processes so that they can be allowed to react directly to what is being experienced. This can perhaps be compared to playing a piano: if you wish to play the air from Beethoven's Ninth Symphony, you must discipline your mind to read the notes and your fingers to play the keys; only when you are proficient at these actions can you respond directly to the original piece and play it as Beethoven meant it to be played. Then, Beethoven's music virtually plays itself through you, though of course there are still countless possibilities of interpretation. Similarly, meditative thinking enables the landscape to think itself through you, and there is no one right way to do this. Many forms of interpretation, in other words, are possible. What happens in calculative thinking is that the pianist imposes a tempo and phrasing that have nothing to do with the original, syncopating it perhaps. More drastically, and this would be calculative in its most scientific form, the music is ignored and the piano is taken apart to find out how it works.

All of this gives no more than a rough outline of the nature of meditative thinking, and to prevent gross misunderstandings, I will have to qualify my discussion in several ways. First, Heidegger's philosphy is a profound and intricate inquiry into *Being*--literally how things are, the facts of existence. This

---

8. Martin Heidegger, "Memorial Address," in *Discourse on Thinking* (New York: Harper and Row, 1966), pp. 46ff.

inquiry has its roots in pre-Socratic reflection, addresses most of the enduring questions of Western philosophy, and proposes that "man is not the Lord of beings; he is the shepherd of Being."[9] This means, to put it briefly and at the danger of distortion, that human responsibility lies not in seeking to dominate and control nature and human nature, but in accepting the role of guardianship or stewardship of everything that is. This responsibility is a concomitant of both self-consciousness and language, neither of which we have chosen to possess. These qualities do not make humans superior to plants, animals and matter, as Descartes and his fellow rationalists would have us believe, but comprise a sort of gift that should be used to tend Being in all its forms, much as a gardener tends his garden or a mother tends her baby with thoroughness and care.

A second qualification is that Heidegger's philosophy cannot be regarded as some simple, alternative social-science methodology that can be quickly learned and used as occasions demand. It is a deep pool of ideas and arguments that takes several years to begin to comprehend, and which on rereading always offers new insights. The language Heidegger uses and the form of his arguments are not commonplace and one has to persevere with them. Yet having once grasped the gist of Heidegger's philosophy, it opens many possibilities--not, of course, in the sense in which logical positivism enhances the scientific procedures used in laboratories that can lead to new technologies, products and policies, but in the way one conducts one's thinking and being. A third qualification about appropriate thinking that should by now be obvious, but which I will mention for emphasis, is that there is no simple set of rules or procedures than can be learned and applied in order to get guaranteed results. Appropriate thinking requires a continuous effort *not* to impose arbitrary *a priori* ideas and to let things be themselves.

I do not use Heidegger's philosophy in a very deliberate way in my investigations of landscape and place. Rather, I have tried to understand and absorb his thinking, and I hope that this reemerges in a not-too-distorted form in my own writing. Sometimes, of course, this re-emergence is self-conscious; often, however, it seems to be unself-conscious, and only subsequently do I become aware of how much Heidegger's philosophy has colored what I have done and written.

---

9. Martin Heidegger, "Letter on Humanism," in *Martin Heidegger: Basic Writings*, ed. D. F. Krell (New York: Harper and Row, 1977), p. 221.

Describing

The tangible product of clear seeing and meditative thinking is for me a description of a landscape. Description seems to be held in contempt as a lower form of intellectual activity by those who espouse theory and scientific explanation. This opinion is quite misguided. Ludwig Wittgenstein, who with Heidegger ranks as one of the major philosophers of the twentieth century, wrote that "we must do away with all explanation and description alone must take its place."[10] Like all such elliptical remarks, this is not as simple as it seems on first reading. "Description" is not confined to an inventory of objects or a list of facts and measurements. It can mean, for instance, historical or legal interpretations of events. Thus, a prosecuting attorney presents the facts of the case as he understands them, while the defense lawyer presents a different description of those facts. A description can be precise, factual, evocative, imaginative, well or badly written, concise, quantitative, and so on. Someone reading a description may take it to be an argument, but to call it that suggests that a deliberate attempt is being made to be provocative. To the person who has written it, a description is simply the most true and most accurate account of the state of things that could be presented.

In his first book, the *Tractatus Logico-Philosophicus*, Wittgenstein sought to transcribe the realm of what could be said clearly and unambiguously by logic and mathematics, and having done this he concluded that an enormous area of life remained that could only be "shown" or "passed over in silence."[11] So he distinguished two realms of knowledge: the one of "saying" or logical analysis, and the other of "showing" or the intersubjective communication of meaning. These two realms of knowledge are not interchangeable: "What can be shown, cannot be said."[12] The realm of saying corresponds to that of a strictly logical-positivist science seeking to delineate the facts of the world, and Wittgenstein's comments on this point attracted much attention from logical positivists. The realm of showing was, however, more difficult to convey, and Wittgenstein concluded that nothing sensible could be said about it, though he did not by this intend to diminish its importance. By the

---

10. Ludwig Wittgenstein, *Philosophical Investigations* (Oxford: Basil Blackwell, 1974), section 109. Wittgenstein's two essential works are *Philosophical Investigations* and the *Tractatus Logico-Philosophicus* (London: Routledge and Kegan Paul, 1970). Clarifications of these two difficult books are respectively Max Black, *A Companion to Wittgenstein's Tractatus* (Ithaca: Cornell University Press, 1964); and Garth Hallett, *A Companion to Wittgenstein's Philosophical Investigations* (Ithaca: Cornell University Press, 1977).

11. Ludwig Wittgenstein, *Tractatus Logico-Philosophicus,* section 7.

12. Ibid., section 4.1212.

time he had finished writing *Philosophical Investigations* some twenty years later, he had rethought this position and much of this later work can be understood as a statement that meanings can be clarified by attending to and elaborating their contexts. It nevertheless remained impossible for Wittgenstein to say exactly what meaning is. What can be done, he suggests by example, is to think clearly about what one is doing, to be aware of the context, and to describe clearly what there is. In the realm of meaning and of showing--that is, the realm of most of life--logical explanation has but a minor role to play.

Landscape is part of the realm of showing, both literally in that it is primarily a visual phenomenon, and indirectly in that it cannot be exactly defined. Consider what it means "to study a landscape." This could involve sitting and looking at a scene, but this is "studying" only in its most trivial and superficial sense. If studying involves some active mental and physical processes, it might require that, for example, building types be classified and counted, behavior patterns mapped, slopes measured, and so on. Yet this is dealing with buildings, behaviors and slopes rather than with landscape. Landscape presumably means more than this, but just how it is more is difficult to say. If we try to define the upper or lower limits of a landscape we cannot--except in an entirely arbitrary way. Although the horizon seems to be one sort of clear boundary, in fact it moves as we move. Furthermore, though we can see a landscape and show it to others, though we know what it is, it has no firm material substance; it cannot be touched, embraced, measured or even walked around on the outside. Only by a distorting act of imagination can we separate ourselves from a landscape and treat it as an object.

Landscapes are our constant companions, always present yet incapable of precise, logical, objective definition. Nevertheless, with some skill in seeing and in receptive thinking, landscapes can be described. In *Philosophical Investigations,* Wittgenstein wrote in one especially apt section, "I look at the landscape, my gaze ranges over it, I see all sorts of distinct and indistinct movements, *this* impresses itself sharply upon me, *that* is quite hazy...And now look at all that can be meant by 'description of what is seen'."[13] He is telling us that there are countless descriptions of any landscape, descriptions distinguished not simply because they select different features for emphasis but also by their levels of insight and intensity. In each description, there is an

---

13. Wittgenstein, *Philosophical Investigations,* p. 200.

individuality that reflects the biases and attitudes of the author, yet they can all be accounts of somewhere that is recognizable to others.

I do not spend my life, academic or otherwise, seeking out landscapes and problems which can be researched using clear seeing and appropriate thinking and then described. These are methodological elements which I can, by some effort of reflection, identify in my own work, and which I read about in the works of artists and philosophers. They are, I hope, implicated in what I write and describe, but they are unequivocally not the means by which I investigate places. In such investigations--for instance, my present concern with the development of modern landscapes--I simply find out as much as I can from historical, economic, or even statistical sources, make continual observations, make sketches, take photographs, and then try to piece together the bits into a coherent account. There is nothing elegant in this. It involves some fumbling, some luck, and a great deal of perseverance and patience in watching, writing, thinking, and rewriting. The end result is a description of what things are like as I understand them. This description should derive from the subject matter, but it also reflects the philosophical convictions to which I hold and my biases in seeing and watching. The results will not shatter worlds nor change the course of history, yet for me they are far from trivial. John Ruskin thought similarly, for he wrote that "the greatest thing a human soul ever does in this world is to *see* something, and to tell what it *saw* in a plain way . . . To see clearly is poetry, prophecy, and religion all in one."[14]

Conclusion

If I am concerned with behavior in my work, it is chiefly with my own behavior--specifically, my seeing and thinking. By attending to these, I try to write reasonably clear descriptions of what is happening to landscapes. Of course, landscapes are themselves the more or less enduring manifestations of the behavior of the institutions and individuals who create and use them, so perhaps my comments about landscapes have behavioral significance. That I will leave for others to decide; it is not part of my explicit intention to see them in such a light.

In this age of scientific explanation and technology, mere description based on personal efforts of seeing and thinking does strike people as laughably futile. My writing has been described as

---

14. Ruskin, *Modern Painters,* vol. 3, chap. 16, section 28.

"pseudish," "a sermon of nonsense," "reactionary," "lacking a foundation in the material basis of history," "a threat to the intellectual rigor of geography." Such criticisms are no doubt true for someone who understands the world to be filled with technical problems such as how to find optimal locations for banks, or how to use computers to study cognitive configurations. At one level of existence, such problems are important--someone has to decide where the banks and traffic lights and mailboxes are to be placed. However, there are other levels of problems where matters are less technical and more fundamental. The erosion of the distinctiveness of places, the increasing artificiality of everyday life, the long-range implications of the engineered management of environments, the threats posed by nuclear and chemical weaponry are all political and ethical matters which cannot be addressed using conventional technical approaches. In fact, these are problems that derive from the diligent, systematic application of rational methods of inquiry, and their solution clearly transcends the scope of those methods. One cannot continue endlessly to use more science to solve the problems of science, just as more alcohol does not cure alcoholism. In order to come to terms with such profound issues and to formulate a sensible perspective on them, I can see no alternative other than aiming for independent observation, responsible thinking and writing. This way of working offers no swift technical solutions and no clearly defined paths of action to follow. It should, however, help to check unthinking dependence on statistics, Marxist dogma, or *a priori* models of behavior.

CHAPTER 15
PHILOSOPHICAL DIRECTIONS IN ENVIRONMENTAL PERCEPTION AND BEHAVIORAL GEOGRAPHY
A COMMENTARY

Joseph Sonnenfeld

A discussion of the nature of a discipline or sub-discipline generally involves biases, and mine ought to be made explicit. My interests are in rather conventional behavioral and geographic issues, related to understanding (a) geographic landscapes and why we behave toward these as we do; and (b) place and community, in particular why some places and communities succeed, have influence, elicit positive responses, and why others fail, are ignored or elicit negative responses. I have a more specific research interest in problems of adaptation to normal and extreme environments, in terms of both sensory and perceptual adaptation, as well as conscious technological and social adaptation to what it is the environment permits, constrains, or stimulates. I assume that in all cases it is people--ultimately individuals--who are so stimulated or constrained by environment. I have also worked on concepts of the behavioral environment and environmental personality; and I'm currently involved in research on spatial behavior, emphasizing the cerebral as well as environmental and experiential correlates of success in wayfinding.[1]

My approach to research in behavioral geography is to assume the need for techniques that permit comparisons of populations, which generally implies some means for quantifying data; this, in turn, requires the development of instruments and interviewing procedures that can be administered either at the individual or group level. I further assume that if other researchers use my techniques, they will be able to verify my results, or, if they

---

   1. See, for example, Joseph Sonnenfeld, "Multidimensional Measurement of Environmental Personality," in *Spatial Choice and Spatial Behavior,* ed. Reginald G. Golledge and Gerard Rushton (Columbus: Ohio State University Press, 1976), pp. 51-66; "Resource Perceptions and the Security of Subsistence," in *Dimensions of Human Geography,* ed. Karl W. Butzer, Research Paper no. 186 (Chicago: University of Chicago Department of Geography, 1978), pp. 15-24; "Egocentric Perspectives on Geographic Orientation," *Annals of the Association of American Geographers* 72 (March 1982): 68-76.

cannot, that they will raise questions concerning the population sample, the instruments used, or the interpretation of data. I further assume the need for flexibility in research technique, because not only are populations different one from another, but also the individuals who make up these populations tend to differ at some level, not to mention also differences in the environmental settings of the behaviors I am trying to understand. All of which seems to make me what one might refer to as a pragmatic practitioner of positivist persuasion. It is such a naive philosophical perspective which gives me problems with some of the papers I have been asked to critique.

Edward Relph's approach is basically egocentric. His concern is to communicate what he sees in and thinks about specific landscapes, without generalizing and without worrying about justification or verification. The descriptions of landscapes that he communicates are the products of "clear seeing" and "meditative thinking;" he encourages us to resist anyone who would alter the landscape that justifies meditative contemplation, and, as well, resist the efforts of those who invoke devious methods of behavior control to alter environmental perception and priorities. Relph seems to confuse behavioral geography with what Skinnerian behaviorism is all about. He raises suspicions about behavioral control, with the implication being that we have to be on guard against such efforts, and this includes any geography which produces data likely to be used for such control--as if teachers, parents, and friends, among others, are innocent bystanders in the human-conditioning process.

All one can do is agree that it must be nice to live in a world where our main professional concern is "to see," and to describe what we see, and to teach others the value of such seeing; to encourage passive contemplation of what has been seen, and, perhaps in more uncharacteristic activism, attempt to prevent manipulators and developers from changing those landscapes still "out there" to be valued for what some of us as individuals can see in them. The question of why some cannot "see" landscape, or see it so differently one from the other, or why some are frustrated or provoked by what they see, apparently does not get asked. Similarly, why some do and others don't make the effort to obtain understanding through contemplative seeing also seems an irrelevant question to Relph, as it would also have seemed irrelevant, curiously, to B. F. Skinner. I agree with Relph's sense of the complexity of behavior, of place, and of landscape; and of the unlikelihood or either complete explanation or total control of behavior. But he hasn't quite convinced me to give up trying to get more adequate explanation of variability in environmental behavior, nor to abandon the effort to increase public sensitivity to environmental and community quality issues.

Douglas Greenberg's paper offers a sociocentric--structural-Marxist--perspective on environmental behavior. Instead of a concern with the distinctiveness of individual seeing and satisfactions, there is more explicit recognition that what and how we see and how we respond to such seeing is not independent of social and historical contexts, quite apart from how aware we are of the social and historical processes responsible for what it is we see, feel, or contemplate. Plekhanov, who wrote of the importance of individual initiative for influencing historical events, also described economy-determined social relationships as having their own "inherent logic," such as causing people to "feel, think, and act in a given way and no other."[2] It is evidence of progress that Greenberg, some eighty-five years later, can speak not only of the continuing struggle between classes and between ideologies, but also of man's reflective self-interpreting nature, of mutual concerns for "social reproduction," and of the importance of affective relationships; and in ways which transcend narrower concerns with economic and social structures.

I especially like the concept of "structuration," which views social structures as enabling of behaviors critical for "social reproduction," though there is obviously much still within society to coerce and condition appropriate social behaviors. The concept of "rule-governed creativity" is also provocative, if unsurprising in a Marxist context. But in broader perspective, too much seems to get missed in the Marxist's focus on Western and Western-exploited societies, and on the human nature that capitalism ostensibly produces. Much which is important for understanding community and environmental behaviors gets ignored, for example, such as relates to ethical, religious, or supernatural belief, not to mention also those behaviors which derive from a play drive and aesthetic sensitivities. That the economic system has influence on religion, play and aesthetics does not mean that it is able to explain why these latter are as important as they are, nor as variable as they appear to be within any given social class or environmental setting.

---

2. George Plekhanov, *The Role of the Individual in History* (New York: International Publishers, 1940), p. 60.

Equally problematic are some rather fundamental questions relating to priorities. How does one justify, from a Marxist perspective, programs designed to protect the quality of a specific environment or landscape when the well-being of a community dependent on that environment may be adversely affected as a consequence? Similarly, how does one decide between short-term and long-term social or environmental goals when these conflict? And is it well-being of the local, regional, national, or global community that Marxists consider the most critical when making decisions on environmental issues? Yet, I'm uneasy with some of these questions, since they imply that there are good and universally acceptable measures of environmental and social quality, and that one can be objective in assigning priorities when conflicts between community and environment arise, neither of which is obviously the case.

The two remaining papers of this series raise meatier methodological questions. Seamon's presentation of the phenomenological perspective seems to call positivist science to task for being too simple-minded, and too slow or cautious for dealing with phenomena that are generally more complex than controlled positivist approaches permit treatment of, given the role of the affective and experiential in much of the environmental behavior of interest to geographers. Yet, as with Marxist approaches to behavioral analysis, there is apparently no one correct way of "doing phenomenology." And since results of phenomenological inquiry are likely to be specific to the investigator, it is unclear how the objective observer is to produce the "accurate qualitative descriptions which will provide a basis for authentic conceptual portrayals of the various dimensions of the person-environment relationship" that phenomenologists seek. In short, how does one verify the authenticity of the conceptual portrayals that result? "Intersubjective corroboration" may be possible in certain restricted circumstances, but this possibility begs the question of how one chooses the corroborators.

Perhaps as bothersome as the problem of verification is the assumption that the general experiential structures, patterns, and essences that the phenomenologist seeks are beyond the reach of positivist inquiry, or at least beyond the kind of theorizing that is needed to guide positivist research. The image projected is that of positivists with paradigmatic blinders who are incapable of seeing the structures, patterns, and essences that the more empathetic phenomenologists are able to identify in the behaviors and settings of the places they study. Equally curious is the suggestion that the phenomenologist's goal is not explanation but an understanding more productive of empathy. If the purpose of such a distinction is to avoid the implied sterility of reductionist explanations, it also raises questions as to the usefulness of the empathy that is generated if explanation is not part of the understanding that is achieved. Reference to Norberg-Schulz's work on the sources of environmental meaning doesn't help to clarify this distinction between explanation and understanding when environmental meaning is described as deriving from the "qualities of natural environment meeting together in place," to yield a given "experiential 'feel'."[3] Is this an explanation of environmental meaning, or only a more acceptable phenomenological style of reductionist understanding?

Efforts to distinguish between phenomenological and positivist approaches seem to represent a quibbling that I do not think does justice to the contributions of either positivist or phenomenologist. Practitioners of both appear to be concerned with objectivity, and also seem to recognize the potentials and limitations of observer bias and experience for understanding human behavior. As an example, Seamon's phenomenological work on "place ballets"[4] and Roger Barker's work on "behavioral settings"[5] clearly overlap in research concerns, despite their differences in assumptions and methodology.

The essay by Reginald Golledge and Helen Couclelis on human spatial behavior--a positivist perspective--is the one I'm most comfortable with, except for the fact that it seems to define the issues of modern behavioral geography too narrowly, as if these were all related in some way to human spatial behavior. More critically, the model of analytic science that they present appears excessively to limit the mode of acceptable scientific discourse and inquiry, at least from the perspective of one who feels comfortable with much of the science practiced by the non-quantitative variety of anthropologist, sociologist, and psychologist. One has the sense, in their form of positivism, of a geography concerned with the sterile extraction of cognitive structures from human image-conveyors, and of preference structures from multiple-checklist inventories and multidimensional-scaling output.

---

3. Christina Norberg-Schulz, *Genius Loci: Towards a Phenomenology of Architecture* (New York: Rizzoli, 1980).

4. David Seamon and Christian Nordin, "Marketplace as Place Ballet: A Swedish Example," *Landscape* 24 (1980):35-41.

5. Roger G. Barker, *Ecological Psychology* (Palo Alto: Stanford University Press, 1968).

Ignored here are the rather basic issues of personality and motivating experience, and the relationship of these to level of competence in performance of environmental tasks.

If the Golledge-Couclelis concept of analytical behavioral geography seems unnecessarily narrow, Seamon's criticism of the positivist position seems equally to ignore positivist potentials. It is not as if the phenomenologists are the only ones to conceive of indissolvable unities; nor should the existence of such unities be assumed, short of evidence for the inadequacy of reductionist explanations of variability in place-behavior relationships that concern the positivists as well as phenomenologists. Nor is there reason to assume that a positivist perspective necessarily ignores the complexities of experience, motivation, and setting. That the positivists at times find these factors difficult to work with does not mean that they are considered non-identifiable and non-explainable. It may only be that the more analytic of "conventional scientists" among behavioral geographers have gone beyond the concerns of phenomenologists in realizing that there are prior questions that need resolving before probing into the intricacies of personal experience, the values which motivate or generate behavior, and related complexities of sense of place and of community.

One could appropriately apply the structuration concept discussed by Greenberg to the phenomenologist-positivist controversy and view both perspectives as enabling, each capable of providing the other with insights, direction, and ultimately the means to verification that proponents of both approaches seek. The phenomenological perspective, in this sense, has a dual role: in part, it is pre-scientific in its treatment of the experiential and value-laden dimensions of environmental behaviors relating to infinitely variable places and landscapes. It is also post-positivist in its treatment of the incomplete products of reductionist analytic research, which often tends to be limited by the probabilistic nature of most human behavior, not to mention also the practical limits on the range of setttings in which such behavior is studied. There are only so many people in so many situations that one is able to contend with experimentally. It is not that the positivist has no interest in the less commensurable behaviors; rather, the problem is more often one of professional priorities. Phenomenologists would probably feel better about what positivists do were the more curious among these latter not always getting "sidetracked" into probing the "whys" of their findings, and were they to appear less deliberate and non-commital in the ways that the positivist stereotype is normally portrayed.

Consistent with the position expressed above, Seamon makes a plea for behavioral research which brings different insights together, instead of that which produces the philosophical differentiation exhibited by the papers presented in this series. Perhaps part of the problem is in the nature of geography as a discipline. Not only are geographers synthesizers in regional and global terms, but equally we have a tradition for emphasizing differences within and between places. While a focus on such differences may in fact help uncover critical underlying processes that have broader explanatory value, some of us do not get beyond the difference-seeking mode. This asymmetry in analytic style also characterizes the research that we do and why we do it; we tend to emphasize not what our research contributes to geographers and other scientists who have related concerns, but rather how our work differs from what others do.

The point has been made that scientists tend to think in terms of similarities, by contrast with others--in particular those in administrative or decision-making positions--who tend to think in terms of differences.[6] Geographers seem to fit both groups, but it is unclear whether that which we prefer is based on training received in a particular type of geography, or whether there is something more basic in the subculture or personality of geographers that conditions us for operating in one or other mode: difference seeking versus similarity seeking. Which mode we favor may also be related to distinctive cognitive styles--holistic, visually descriptive, and spatially-sensitive at the one extreme; experimental, analytical, and abstract at the other.[7] I might only suggest that most of us retain a neural (cerebral and cognitive) plasticity still; we are not bound by thinking styles adopted during our professional schooling or beyond. And while it may be difficult for some to give up ways of thinking that seem to have worked well for problems we have dealt with in the past, the problems that still bother positivists and phenomenologists might be better resolved if each would recognize what the other has to offer, rather than what it is that separates them. The realities of environmental behaviors and of the settings of these behaviors justify a more complex

---

6. J. P. Maxfield, "The Executive vs. the Scientist," *PMR-MP-59* (Point Mugu, California: Pacific Missile Range, n.d.).

7. Joseph Sonnenfeld, "The Communication of Environmental Meaning: Hemispheres in Conflict," in *Nonverbal Communication Today: Current Research*, ed. Mary Ritchie Key (Berlin: Mouton Publishers, 1982), pp. 17-29.

approach to behavioral research than either group by itself seems able to present.

In terms of scope, all the papers appear deficient in certain respects, and curiously narrow by "conventional" geographic standards. For example, none deals explicitly with environmentalist issues, or with themes of cultural or technological diffusion and adaptation. None deals with possible genetic influences on behavior, nor with the implications of developmental factors and early conditioning for our understanding of place preferences and sensitivities. None really deals with the immediacy of the problems that behavioral geographers are or ought to be concerned about, or with the contribution that we are able to make, or have made, toward resolving critical behavioral and environmental problems.

It is not especially clear from these papers what the purpose of research in behavioral geography should be. Is it our concern to promote empathy? To inform others of the richness of the world's culturally distinctive environmental behaviors and relationships? To find more effective means for controlling inappropriate environmental behaviors? To facilitate access to work, shopping, or recreational places by those with variably developed spatial skills? Or is it, as with much of science, to obtain explanation and understanding of the differences and similarities from place to place, quite apart from applications?

Realities often interfere with the expression of professional values, and sometimes it is difficult to distinguish between values which are professional and those which are personal. How do we justify imposing these values on others, or on their working and leisure environments? What do we gain by increasing the level of public sensitivity to the importance of certain qualities of the geographical environment? Is there nothing of equal importance that may be lost by being so sensitized? What is the balance that we seek between well-being of community or society and well-being of environment? In terms of research methodology, how do we justify the depth to which at least some of us feel it is necessary to probe the personal sources of environmental behaviors? This is an ethical issue that cross-cultural psychologists, among others, have seen need to consider.[8] As a group, cross-cultural psychologists are more active in field work than other psychologists, which may be why they were able to agree that they had an obligation to consider as primary the welfare of those whom they studied. Can we justify

---

8. *IAACP Cross-Cultural Psychology Newsletter* 11, nos. 1-5 (1977), ed. Frances E. Aboud and Donald M. Taylor (Montreal: McGill University, Department of Psychology).

otherwise the intrusion that our research in behavioral geography often requires into the private lives of individuals and communities? Are the benefits of our research such as to allow us to ignore the fact that we intrude? And at what cost to understanding if we decide to avoid research which seems excessively to intrude on respondent privacy and community good-will?

This issue of research ethics deserves much more attention than it has received from geographers of the behavioral or other traditions. At this point in the development of behavioral geography, the question of ethics and related questions of purpose seem at least as important for us to consider as the issue of what we are or are not able to do with the research methodologies available to us.

CHAPTER 16
MAPPING THE MOVEMENT OF GEOGRAPHICAL INQUIRY: A COMMENTARY

Robert Mugerauer

A pattern clearly emerges in these papers on philosophical directions in behavioral geography: (1) the authors develop a position on one or the other side of the "classical" dichotomy of the subjective versus objective modes of knowing; (2) they acknowledge the failure of an extreme or rigorous version of the side of their initial choice; (3) they attempt to move to a more defensible middle way; (4) they wind up more or less reaffirming some version of the perspective with which they started. As a result, the papers are old maps which show where geographers are today. But the surprising agreement on what is needed in geography and the openness of the attempt to arrive there--step 3--indicate that there is new, uncharted territory ahead for which we are just beginning to delineate our maps. Here we see both the continuity with and progress beyond the work done fifteen years ago.[1]

Apparent and Old Directions

On a first reading, it appears that geography is polarized by the "classical" dichotomy indicated above. For example, the essays of Golledge, Relph, and Greenberg do indeed tend to identify their approaches and allegiances as positivist or structural versus subjective or existentialist. Thus, the very topic of behavioral geography gets temporarily set aside as distinctions are worked out. For example, is 'behavioral' as used by Relph understood according to behaviorism, and, accordingly, as a species of scientism and logical positivism? If so, then the opposite view would seem to be that of existentialism or phenomenology. With the lines so drawn, it seems that geographers are repeating the philosophical quarrel of the modern era, and especially of the current century, about what science is. This is of the highest importance, since the foundations of geography as a science and the validity of its approaches are in question here. Such issues are, at least logically, prior to concrete analysis.

---

1. David Lowenthal, ed., *Environmental Perception and Behavior,* University of Chicago Department of Geography Research Paper no. 109 (Chicago: University of Chicago, Department of Geography, 1967).

Clearly, the opposites line up in a familiar way: positivism versus phenomenology; behaviorism versus existentialism; empirical scientific approaches versus more personalistic methods. These categories are "classical" in two ways. First, and historically, we have the long differing schools of methodology and epistemology. Specifically, the major arguments on behalf of scientific method in our century were posed by logical positivism in its various forms,[2] and countered by phenomenology and existentialism.[3] Second, in the debate between these two camps, certain arguments and counter-arguments were developed and refined--to everyone's credit, often by the self-scrutiny of each camp. Thus, for example, it was first argued that for a proposition to be true it had to be verified; then, recognizing the limitations of empirical reach, the statement was modified to include verification in principle; finally, as a foundation for methodology, the criteria crumbled under the burden of self-reference: it could not be verified that for a proposition to be true it had to be verifiable in principle.

Not to belabor the point or reconstruct the intellectual history of the last sixty years, the arguments back and forth became polished enough to amount to "moves," like plays in chess. To some degree, the contributers here, especially Golledge and Seamon, repeat these moves. This process is not without value. At the same time, while practitioners continue to debate theory, theoreticians generally agree that any sort of strict positivism is untenable. This conclusion is why positivism and strict transcendental phenomenology have moved over for new positions, e.g., Polanyi's post-critical work, hermeneutics, structuralism, deconstruction, and so on. In other words, the "classical" opposition is somewhat *passe*. That does not mean it is untrue or otherwise unimportant.

What can be learned from this development is that behind the various specific polarizations exhibited in the papers of Golledge, Relph, and Greenberg, for example, we can find the more fundamental distinction between objectivism and subjectivism. On one hand, we have the claim that in order to be a science, geography must have a method which yields objective results. To be objective requires that the data or facts stand free from subjective coloration or

---

[2]. Robert Ammerman, *Classics of Analytic Philosophy* (New York: McGraw Hill, 1965); Morris Weitz, *Twentieth Century Philosophy: The Analytic Tradition* (New York: The Free Press, 1966).

[3]. Edmond Husserl, *Logical Investigations* (New York: Humanities Press, 1970); Martin Heidegger, *Being and Time* (New York: Harper and Row, 1962); Mat Scheler, *Selected Philosophical Essays* (Evanston: Northwestern University Press, 1973); Herbert Spiegelberg, *The Phenomenological Movement* (The Hague: Martinus Nijhoff, 1960).

distortion. Indeed, this is the model of positive science: to be free of values, situations, and personal idiosyncracy. On the other hand, the point is made that human knowledge of what humans experience is, like it or not, subjectivistic. To be a geographer is to be human and therefore enmeshed in subjective viewpoints. As noted, Golledge and Greenberg repeat now classical moves and countermoves. This repetition helps us keep hard won insights. In addition, Golledge and Greenberg play out the long tradition of subjectivism versus objectivism which is our heritage since Descartes formalized the distinction.

Golledge, by his geneology, derives from Descartes, Hume, and Comte, while Relph is akin to Sartre. The result is that value-free objective knowledge is opposed to value-laden expression and perception; object is opposed to subject because the subject is seen as the perceiver of the object which is perceived, and the object is defined as what stands over against the perceiving subject. Now, one major problem which should stop us here is that this opposition itself is untenable. Both sides of the "classical" difference agree to the assumptions behind the difference and merely choose sides. Since the Cartesian position, however, metaphysically and epistemologically separates what in fact are only facets of complex integrated reality, the Cartesian dualism is untenable. Thus, both the subjective and objective positions are "overcome" in the unitary attempt at understanding found in the work of Polanyi, Heidegger, and scholars doing hermeneutics, critical theory, and non-subjective phenomenology.[4]

Greenberg attempts to develop the relationship between structure and agency in order to move between or beyond Golledge's interest in logical-mathematical structure and Relph's holding on to his own perceptions. Greenberg would have us see that Marxism is a mediating position--the synthesis, as it were, of the thesis and antithesis. This argument, in fact, does not work. Actually, both the subjective and objective camps have attempted to co-opt Marxism in recent years. Sartre, for example, attempted to show that existentialism and Marxism were both humanistic and ultimately subjectivistic; other Marxists reject this argument, stressing that Marxism derives from a most rigorous form of the objective-historical dialectical materialism.[5] Hence, although

---

[4]. Michael Polanyi, *Personal Knowledge* (Chicago: University of Chicago Press, 1958); Martin Heidegger, *The Question Concerning Technology* (New York: Harper and Row, 1977); Josef Bleicher, *Contemporary Hermeneutics* (London: Routledge & Kegan Paul, 1970); Hans-George Gadamer, *Truth and Method* (New York: Seabury, 1975); Robert Mugerauer, "Concerning Regional Geography as a Hermeneutical Discipline," *Geographische Zeitschrift* 69 (Spring 1981): 57-67.

Greenberg stresses structure with Marx and Althusser, the Marxist rebuff of Sartre stands. In Marxism, positive science still dominates. That is not to say that Marxism and positivism are identical; but, in its theoretical foundation and its *praxis*, Marxism also belongs on the objective side of the Cartesian dichotomy. It is no synthesis, but a part of what yet is fractured.

What do these conclusions mean? That Golledge, Relph, and Greenberg get us nowhere? Certainly not. Rather, much of what they say reiterates the classical polarizations and thereby forces us to think them through again. Here, one sees the continuation of debate taken up at the 1965 Columbus symposium with the contrasting presentations of Yi-Fu Tuan, Robert Beck, Joseph Sonnenfeld, Robert Kates, and Kevin Lynch. Such thinking through again is never done too often if we would be thoughtful. Recall that Heidegger defines thinking as remembering thinking.[6] In addition, this style of thinking helps us to move beyond the already familiar differences. The result is that we recall necessary background and arguments in order to move forward.

New Directions

Golledge, Relph, and Greenberg *appear* to move in various directions. But, if the basis of their positions is the *apparently* valid dichotomy of subject and object, then their movement is not what we or they might suppose. That is, the old map of subjective-objective territory still does work to see some general directions in which geographers have and will continue to move, especially since, as in all disciplines, faculties continue to teach, base research on, and believe what they learned in graduate school, regardless of the current view. Presently, since there is a lag between understanding and action, the political power of departments remains with positivists or is "shared" between positivists and other points of view.

But, and this must be stressed, there is more agreement between Golledge, Relph, and Greenberg than is apparent, largely because they are thoughtful people who work concretely in the discipline and because, I sense, the pressure of reality informs what they really hold and do more than their "extreme" theoretical positions do. If we listen carefully, we find that their papers speak with two voices. For example, Golledge presents the critical self-destruction of positivism, looks beyond it, then finally tries

---

5. Cf. George Novack, *Existentialism Versus Marxism* (New York: Dell, 1973).

6. Martin Heidegger, *What is Called Thinking*? (New York: Harper and Row, 1968).

to reconstruct an analytic mode of discourse which emphasizes mathematical structure, verification of knowledge, and unambiguous propositions. Relph overdoes the subjective at the start of his paper, proceeds to a more moderate, non-idiosyncratic understanding and insight, then undoes this balance by reaffirming his own behavior and seeing. Greenberg critiques the voluntaristic and deterministic alternatives, examines structuration as a middle way, but still ends with a dichotomy. Each argument begins and ends in a classical position or variation, but along the way moves to new territory, even if withdrawal comes too quickly.

These two voices, classical and original, articulate, respectively, what the contributors already have and what they want. What they personally hold is that geography requires a way to make sense out of the environment in which we live, an environment both natural and humanly constructed. What the contributors mean is that, as a community, we need to make sense out of this environment and, by way of self-awareness, come to an understanding of our place in it. Hence, a geographer must see clearly and deeply, share what is seen in some sort of systematic manner, but *not* be epistemologically naive or aberrant. This communal seeing and non-idiosyncratic learning is what we call science. Again, Golledge, Relph, and Greenberg often speak in extreme subjectivistic or objectivistic ways; but, on the whole, as full people and geographers, none of them finally seems to hold to these extremes.

Consider a specific example of what they want to develop. All sides agree that the need is for *knowledge*. Knowing how things are and why drives inquiry. It is not this or that methodology that matters in itself. Rather, the methodology is articulated so that we can more consistently and coherently know what lies before us. Method aims at knowing. Since method only serves or leads to knowing, however, method itself is not what we hold on to at all costs. We use the methods which best help us to know. In fact, as Golledge explicitly says, reality is so complex and multi-faceted that we need multiple methods to understand it.

In addition, Golledge, Relph, and Greenberg acknowledge that any methods are humanly informed. Still, to be humanly informed does not imply any radical subjectivism. While knowledge is personal, it need not be idiosyncratic. This recognition is the way beyond Cartesian subject-object dualisms. What we need is a method which enables the human thinker to know the environment without objectifying it or reducing himself to the status of the merely subjective. This aim of study is clearly most difficult: to keep out of our own way as we try to know.

Here, too, Golledge, Relph, and Greenberg admit this realistic, practical goal when they finally do speak in a voice closest to how they actually behave: what matters is not what they are driven to by the argument, but what does the job. In this sense, the three essays are not so far apart. All three writers show a final interest in what works. I conclude that their interest in what does work is very close to the mainstream of American pragmatism. This means that they do not follow the unhelpful conceptual maps, insofar as those maps get us more and more lost in the subjective or objective wastelands. Instead, the three authors attempt to draw new maps with their original ideas. That is what Golledge, Relph, and Greenberg finally show us: to reach our destination, we can go a good way with the old "classical" maps, but now we are sketching and elaborating the maps for what lies ahead. This is why all three contributors stress the modes of inquiry which are open to new seeing and thinking.

There is, as Seamon points out, a way into the new territory which already is charted by phenomenology, insofar as it avoids the Charibdis and Skylla of objectivism and subjectivism. I would argue that a hermeneutical perspective can take us even further. Here, "hermeneutical" can be taken as an extension of the phenomenological--a development which emphasizes the power of language and consideration of historical changes in fundamental assumptions and understanding. Theoretically, phenomenology has its moves and tradition which have scarcely been tried out. For historical reasons, Americans have not yet worked through all the basic moves of phenomenology. This is especially so of the physical and social sciences, but is also true in general, as even philosophy is only now uncovering nonsubjective phenomenology and overcoming the subjective which was taken to *be* phenomenology itself, but is not. In addition, phenomenology has scarcely been carried out as concrete analysis or description--the very point of a method which aims at providing a concrete way of interpreting that which presents itself. The reason for this is that phenomenology is an art, which like the arts of statistical analysis, cartography, and model building can only be learned by continual practice, with others, under the tutelege of a master. Where so few have learned the art of phenomenology even in the core area of philosophy itself, it is not surprising that comparatively few psychologists, for example, and fewer geographers, practice the art. But it is afoot.

We know the phenomenological method is fruitful, as is witnessed by the concrete work in philosophy, the life sciences, architectural criticism, art history, psychology, and the social sciences. That phenomenology can be first-rate science is shown in the work of the scholars and researchers engaged in "philosophical anthropology," a viewpoint and systematic method for doing a multi-disciplinary interpretation of the human and natural worlds.[7] Part of the reason for the success of phenomenology and hermeneutics is that they assume that person and world are given together, both in our prereflective experience and in our critically self-reflexive knowing. What matters, then, is not how subject and object can be independently valid and yet related, but how that which shows itself as it is (the phenomena or reality) is to be most deeply interpreted by us, collectively and over time, so that the nature of what is emerges rather than remains hidden. Phenomenology aims at a method of moving beyond mere particularity, for science must have general insight into the essential nature and patterns of phenomena. Yet, as Relph and Seamon indicate, it does so without reductive abstraction or homogeneous universalization.

As noted earlier, one difference between phenomenology and hermeneutics is that the latter goes much further to stress the importance of language in our interpretation of reality. Since we do, in fact, unavoidably interpret the world to which we belong, it is most important to see how language enables the environment to unfold. This recognition holds that neither language creates reality (subjectivism) nor that language obscures some pure, objective reality (objectivism). Rather, the argument is that, in fact, world already and always is given to us as languaged creatures.[8] Golledge, Greenberg, Relph, and Seamon all point to the need for such a position where the self-critical occurrence of world, embodied person, and language are all seen together. Understanding the environment, then, must involve an understanding of how interpretation arises from and changes with the language we have.

Geographers have come a long way in fifteen years and stand ready to move in new directions. At least several areas to be explored are at hand. As geographers pass beyond subjectivism and objectivism, by whatever method, it is increasingly clear that they will not find new Cartesian permutations of space; rather, what

---

7. Helmuth Plessner, *Philosophische Anthropologie* (Frankfurt am Main: S. Fischer, 1970); Adolf Portmann, *Animal Forms and Patterns* (New York: Schocken Press, 1967); Erwin Straus, *Phenomenological Psychology* (New York: Basic Books, 1966); Marjorie Grene, *Approaches to a Philosophical Biology* (New York: Basic Books, 1968).

8. Martin Heidegger, *On the Way to Language* (New York: Harper and Row, 1971).

awaits is the interpretation of *place,* especially as Heidegger indicated. In the works under construction here, Seamon refers to landscape, place, and locality; Greenberg, to studies of locality, geographically bounded contexts, and reconstituted regional geography; Relph, to the importance of heterogeneous and specific landscapes. The direction is obvious. Places, understood as particularly concrete emergences of world interpreted by humans, will come into greater and greater focus, for example, as we investigate specific patterns of belonging-together which have succeeded and failed.[9] In fact, this theme is a genuinely traditional area of geography, since long before the modern era the earliest works focused on place. In interpreting place in the future, focus will need be given to both *landscape* and human *dwelling* and to the way the two are related. To understand how humans have, do, and can inhabit the earth is indeed the special subject matter of the discipline known as geography.

Further, given this subject matter, geography will also have to turn new attention to *regions,* since regions are a way of thinking about place instead of space. That is, because a genuine region is a heterogeneous place, any conceptual schema which can be systematically applied to the corresponding coordinates in homogeneous space does not yield understanding about the region in any but the most abstract sense. Positively, for a region to be understood as such, the physical and cultural landscapes must be seen as given together, inextricably woven in what might be termed a particular style or local character.

Importantly, when any interpretation of a region begins, we always are already given both its intertwined physical and cultural aspects. This fact means that any interpretation begins with how the region is already articulated, especially in its own language. Language displays the characteristics of a region because it is through local language that regional features first emerge in the lives of those dwelling in the region. This primary manifestation of the unfolding of landscape is the foundation for the geographical investigation which, logically and temporally, is secondary. In short: region must be approached *via* regional language, that is, *dialect.*

---

9. Martin Heidegger, "Kunst and Raum," *Die Kunst and Der Raum* (St. Gallen: Erker-Verlag, 1969); Martin Heidegger, *Poetry, Language, Thought* (New York: Harper and Row, 1971); Christian Norberg-Schulz, *Genius Loci* (New York: Rizzoli, 1980); Edward Relph, *Place and Placelessness* (London: Pion, 1976); David Seamon, *A Geography of the Lifeworld* (London: Croom Helm, 199); Robert Mugerauer, loc. cit.

It should be clear that such a trend in geography is no mere return to old-style dialectic geography or any traditional analysis involving mapping of names, language usage, change, and so on. Rather, a new geography of language is emerging which aims to interpret how, for a given region, physical, cultural, and dialect landscapes are all given together.[10] Or, put another way, the subject matter is the way in which particular regions emerge and are sustained as regions through the languages which bring forth and hold the physical and cultural dimensions together. Instead of the traditional determination and mapping of changing language use, the new geography of language will investigate how language enables regional landscape to emerge and how language succeeds and fails in holding fast the essential features of regional landscape. Place, landscape, dwelling, regions, dialects: these are the new directions geographers are going. Here, too, geographers have begun. Now the task remains to stay underway on these new directions in geography, directions of vitality.

---

10. Robert Mugerauer, "Language and the Emergence of Environment," in *Dwelling, Place and Environment,* ed. David Seamon and Robert Mugerauer (The Hague: Martinus-Nijhoff, 1984).

CHAPTER 17
MAINTAINING THE FUNDAMENTAL PRINCIPLES
A COMMENTARY

Gerard Rushton

The philosophical directions espoused by Relph, Seamon and Greenberg are incompatible with the objectives, methods and procedures of behavioral geography, as I understand the field. There is a history to the usage of 'behavioral geography' and in my view, that history deserves to be respected. At a time when 'human,' 'cultural' and 'social' geography are labels often used interchangeably, the directions described in these three papers will serve only to add the term 'behavioral' to this triad. Other researchers have written about behavioral geography and there is far from unanimity of definition.[1] There is still less unanimity about the measure of value of what has or what might eventually be accomplished by scholars in this field.[2] These three papers extend past usages of the definition in a manner that, if accepted, would destroy the core of the definition of the field as originally defined.[3] I share Couclelis and Golledge's misgivings that many principles, about which so much effort was expended in geography twenty years ago to gain their acceptance and to gain agreement on their value, are being rather too easily discarded in these three papers.

These principles placed great emphasis on the distinction between personal 'knowledge' and 'knowledge' that could be communicated using operationally defined entities in circumstances that, in principle at least, could be replaced. Behavioral geography then was synonymous with a scientific approach. Although

---

[1]. R. G. Golledge, L. A. Brown and F. Williamson, "Behavioral Approaches in Geography: An Overview," *Australian Geographer* 12 (1972): 59-79; R. G. Golledge, "A Practitioner's View of Behavioral Research in Geography," *Environment and Planning A* 13 (1981): 1-6.

[2]. Trudi E. Bunting and L. Guelke, "Behavioral and Perception Geography: A Critical Appraisal," *Annals of the Association of American Geographers* 69 (1979): 448-462.

[3]. K. R. Cox and R. G. Golledge, eds., *Behavioral Problems in Geography: A Symposium*, Northwestern University Department of Geography Studies in Geography, no. 17 (Evanston, Ill: Northwestern University, Department of Geography, 1969).

different methods were proposed,[4] most researchers agreed that it was desirable to observe the subject in a manner that others could replicate, to sample a population using established procedures, to develop tests of the validity of any proposition, and to let data be the judge of the worth or worthlessness of alternative propositions. Such work brought scientific rigor, formerly lacking in this area of human geography, and its practitioners coined the phrase 'behavioral geography' for their work.

The work of Seamon and Relph differs from this tradition in two important respects. Their questions and aims are different and the criteria they would use to decide validity are different. One aim of their research is to gain insights into the processes at work and thus gain understanding of the processes that create landscapes. Another aim is to appreciate the complexity of life and to understand the limitations of any explanations offered by the theories and models of other schools of thought. For Seamon and Relph, I think it is "a matter of connecting action to its sense rather than behavior to its determinants."[5] The role of environmental information here is evidently different from the definition in conventional behavioral geography. What to Golledge is experience, to Seamon and Relph is 'experiential.' I am uneasy with 'experiential' because it emphasizes the effect of the experience on me. Though I do not deny that such a viewpoint might lead me to important insights, I am at a loss to know how anyone else but me would ever know that I had this knowledge, just as I am at a loss to know how I would ever know that Seamon and Relph had attained the knowledge they seek. In reading their papers, I have the impression that this is not an important question to them.

Verifiability is a second principle of behavioral geography of twenty years ago. It appears to have become lost in three of these papers in a sea of '-isms' that behavioral geography is in danger of becoming if it follows the directions of these papers. Couclelis and Golledge write about an insistence on open, public, and intersubjective tests of knowledge by continuous reference to experience. Yet Seamon makes a virtue of the fact that "there are no external props like statistics" or "legitimacy requirements to guarantee the accuracy of the process." Relph quotes Wittgenstein: "we must do away with all explanation, and description alone must take its place." I cannot accept the alternatives that are being

---

4. Ibid.

5. Clifford Geertz, "Blurred Genres: The Refiguration of Social Thought," *American Scholar* 49 (1980): 178.

offered in these papers. Relph counters his criticism that "ongoing attempts at such explanation lead to crass simplification" with a model of seeing, thinking and describing: "for me, this involves looking carefully at specific landscapes, allowing my thinking to respond to the character of these landscapes and then describing them." I see the phenomenological view as possibly providing insights, and I see insights as a first and obviously very essential step to explanation. But explanation and verification are, to me, two sides of the same coin, and in three of these papers I find no sense of the importance of external verification.

A third principle of early work in behavioral geography was that the ability to predict behavior is a test of our understanding. Not only is this principle absent in these papers, but in Relph's work it is specifically rejected as a goal. To Relph, knowledge that could be used to predict behavior is knowledge that could be used to control behavior. We are expected to agree with him that control of behavior is always undesirable. The problem of designing new environments so that defined goals are realized is an old problem generally studied by people who think that their knowledge of human behavior is such that they can accurately predict the behavior that would be found in a new environment. The behavioral basis of design was present in the minds of architects long before it was fashionable to credit individual studies with providing such knowledge. The goal of predicting behavior in new environments is a distinguishing feature of traditional behavioral geography. Schaefer expressed the issue well when, in anticipation of developments in behavioral geography that came about twenty years later, he wrote: "It is our task to make explicit the role these variables play in the social process. In other words, we must try to explore what else would be different in the future if, all other things being equal, the spatial arrangements in the present were different from what they actually are."[6] This question focuses on specific elements of the environment--in this case the geographical arrangement of things--and asks how, as these arrangements change, the behaviors of people change. We wish to know what changes in behavior occur as the geographical arrangement of things change. It is a question that can be answered either by careful observation of behaviors in different environments or, more suitably, by systematically controlling the environments to which people are randomly assigned, and by observing the consequences in terms of

---

6. Fred K. Schaefer, "Exceptionalism in Geography: A Methodological Examination," *Annals of the Association of American Geographers* 43 (1953): 248.

human behavior. It is an important question, because its solution gives understanding and knowledge that, in deference to Seamon and Relph, I must say, we cannot now know that we do not yet know. This knowledge can be used, for good or evil, to control behavior in the sense that it can be used in the design of new environments that will facilitate desired patterns of behavior.

An example of an experience shared by participants at the 1982 AAG Conference here in San Antonio will illustrate my point. Many of those geographers attending have been heard to remark, generally favorably, about the spirit of the city of San Antonio as they have encountered it in the vicinity of the conference hotel where large-scale urban redevelopment efforts in the last ten years have changed the organization of activities in much of the city core area. Those who like it might reflect that it could exist only if man had been willing to manipulate behavior in the sense of designing a rather unique environment centered on the alignments of the old water courses in the downtown area. The developers and planners realized that you and I as users would react to this new environment in a manner that they could predict. The key to the area's development, I assume, was the insight of the people of vision who saw it as a possible place and worked to create its 'genius loci.' Also central to its development, I presume, was a body of knowledge, rooted in experience and explicitly verifiable. In other words, the founders of this downtown development wanted to create a unique environment based on a novel combination of immutable truths about human behavior as known to them at that time. Developing such a body of knowledge in the spirit of "rigor, consistency and unambiguous propositions" is what, to me, the spirit of behavioral geography is all about.

A fourth principle in the traditional view of behavioral geography is that knowledge of how individuals behave in complex environments is sought for its own sake. After finding such knowledge, anyone might, armed with his own ideology, see a use or misuse of such knowledge. Greenberg's paper forsakes this principle in making questions about the role of the political, economic and ideological structures within which people exercise some amount of behavioral discretion. He likes the views of Giddens, in whose work he sees these structures become the forces that either enable certain actions to be made or not.[7] He sees the possibilities for dynamic processes to be at work where certain patterns of behavior

---

7. Anthony Giddens, *Central Problems in Social Theory* (Berkeley: University of California Press, 1979).

both support and perpetuate certain structures which, in turn, become prime forces in restricting or enabling future patterns of behavior. There is no discussion of conditions for equilibrium or disequilibrium (which surely is an important part of such work) nor of the nature of the recursive process by which structure and individuals interact. Greenberg leaves the view that there exists "a critically oriented social science" which, if further developed, might resolve questions of the "motive forces in the creation of landscapes." Greenberg's own ideology appears to be the greatest bar to its development when he concludes by narrowly defining the issue as related to developing knowledge that will change the capacity in human societies for individuals to act as individuals and to exercise their true subjectivity. In defining the issue in this way, attention is diverted to changing the forces that affect behavior rather than understanding precisely how those forces affect behavior in the first place.

Saarinen recently described behavioral geography as "the subfield most noted for its vigorous quantitative theoretical approach."[8] Others have identified philosophical directions that are consistent with such an approach and which, if followed, hold a promise for significant advancement of the field as it has traditionally been defined.[9] The work described in the papers of Greenberg, Relph and Seamon will not advance this approach, and I wish they would choose another label for their work and leave 'behavioral geography' to prosper or decay in the niche developed and cultivated by the group of scholars who share Couclelis and Golledge's acceptance of "the few fundamental principles of analytic discourse": external verification of propositions; validation of explanations, whenever possible, by the prediction of human behavior in new circumstances; and by determining generalizable observations of behavioral response to controlled environmental variability. Elsewhere, these principles have been dismissed as exhibitions of 'instrumentalism' where models are declared to be 'adequate' because they meet their own criteria of goodness of fit to a satisfaction test,[10] but the more basic test, advanced here, is that the explanatory power of models is to be judged by their ability to predict behavior patterns in new environments. Other checks of

---

8. Thomas F. Saarinen and James L. Sell, "Environmental Perception," *Progress in Human Geography* 4 (1980): 531.

9. See Couclelis and Golledge, this volume; and K. R. Cox and R. G. Golledge, eds., *Behavioral Problems in Geography Revisited* (New York: Methuen, 1981).

10. Derek Gregory, *Ideology, Science and Human Geography* (New York: St. Martin's Press, 1978), p. 41.

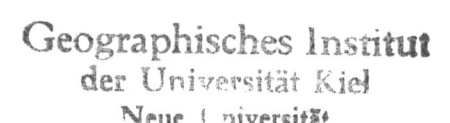

statistical fit are mere bellwether tests to tell the researcher that the route being followed is taking us to this ultimate goal. Avoiding the search for theories, laws and models that can predict human spatial behavior is, in my opinion, a sure recipe for academic demise. If we pursue such a course, our questions will be judged increasingly to be irrelevant and our findings will interest only ourselves.

CHAPTER 18
PERCEPTION IN FOUR KEYS: A COMMENTARY

Anne Buttimer

The invitation to comment on these papers offers a welcome challenge. Many people here remember the enthusiasm and pioneering *elan* of that session on "Environmental Perception and Behavior" at Columbus in 1965.[1] It was my first experience of a national meeting of the Association of American Geographers and I distinctly remember how thrilling it was to find senior members of the profession on a wavelength similar to the one I had come to in my own dissertation work on French geography. There was a puzzle on my mind that day, but I was far too timid to raise it publicly: why, way back in the 1930s, was Hardy's initiative in *La géographie psychologique* so harshly squelched?[2] In this land of boundless opportunity, I told myself, maybe the idea will be judged on its own merits.

Since 1965, a great variety of perspective and style has come to characterize research on environmental perception and behavior in America. Opportunity, of course, is socially contingent, and our perceptions of success or failure are also filtered through cultural lenses. Perhaps the most helpful comment I can offer is an invitation to reflect on the "dream and reality" of the past fifteen years, to look at some of the key philosophical differences among stances articulated in this session, and to speculate on potential common denominators and sources of conflict among them. My own perceptions are no doubt colored by continuing curiosity over the social construction of thought and practice and wishful thinking about dialogue among diverse stances where the integrity of each is respected.

Let me begin my complimenting David Seamon for arranging this session which juxtaposes at least some of the starkest contrasts in style of commitment and research which have come to populate this general field. I suspect that it is from the passion and conviction

---

1. The session was organized by Robert Kates and chaired by Gilbert White; participants included Robert Beck, David Lowenthal, Kevin Lynch, Joseph Sonnenfeld, and Yi-Fu Tuan. Papers in the session were later published as *Environmental Perception and Behavior,* ed. David Lowenthal, University of Chicago Department of Geography Research Paper no. 109, (Chicago: University of Chicago, Department of Geography, 1967).

2. George Hardy, *La géographie psychologique* (Paris: Gallimard, 1939); reviewed by A. Demangeon in *Annales de Géographie* 49 (1940): 134-137.

evident in the expression of each stance, and the animosities evident among them, that the most valuable lessons are to be gleaned. Seamon's introductory paper sets these contrasts in the framework suggested by Gregory for a revitalized human geography; one is invited to develop structural, reflexive, and committed modes of practice.[3] Whether these approaches can be orchestrated or even practised simultaneously may well depend on how adequately we understand the diverse currents of philosophical direction in their own right.

It seems to me that there are at least four distinct schools of thought implicit or explicit in the papers: positivism, Marxism, structuralism, and phenomenology. Each of these perspectives was initially imported from continental Europe to the Anglo-American world, and today's articulation of each no doubt reflects the vicissitudes of the migration and adaptation experience. To understand the roots of those incompatibilities which have been expressed today, or indeed to assess the value of each for the elucidation of environmental perception and behavior, one might well consider the drama of host and guest thought styles in their selective migrations and repatriations across the English Channel and the Atlantic. What is welcomed in the culture, art, cuisine and thought of migrants is that for which the host has already cultivated an appetite; by corollary, the "unheard of" may be misconstrued, rejected, or repressively tolerated, unless there is openness for something entirely new to develop. The style and content of our references to these various "-isms" and "-ologies" today, then, may tell us quite as much about ourselves as it does about the fundamental philosophies themselves.

I use the expression "Anglo-American" deliberately: over half the participants in today's session come with backgrounds in British education and have spent the bulk of their professional careers in a North American context. One recalls that in the pioneering panel there were mostly Americans, nor indeed was there any mention of the Vienna Circle, Marx, Levi-Strauss or Husserl. It is good to have one of those pioneers, Joseph Sonnenfeld, present today, not only to refresh our minds on how the original challenge was perceived, but to let us see what endures of the initial ethos and style of approaching it. The geographers, coming from cultural-historical and man-land traditions, seemed excited over the prospects offered by conceptual and analytical developments in psychology; the

---

3. Derek Gregory, *Ideology, Science and Human Geography* (London: Hutchinson, 1978).

psychologists, coming from psycho-analytic and empiricist traditions, seemed excited about the prospects of considering "spatial" or "environmental" dimensions in their own work.

As these new techniques and ideas were absorbed within geography, the old bifurcation of spatial and man-made traditions reasserted itself; the former, by and large, espousing a behavioral orientation, the latter an ecological one. Ambivalence also grew between those whose primary curiosity was about processes of perception and cognition, and those who were primarily interested in patterns of behavior. What receded from view over the years was that initial goal of seeing connections between the two, e.g., in the context of historical patterns like attitudes toward nature,[4] settlement of the Great Plains,[5] or of problem-solving and policy on flood plain management.[6] From the geographer's perspective, what has been produced may seem like a fragmented array of special insights into particular aspects of human perception and behavior rather than more complete or explanatory descriptions of landscapes.

Such, of course, has been the record in the whole Western tradition, expecially in geography, where the values of empirical inquiry via the "case study" approach have held high esteem. In fact, to have outlined a rational framework in which the special contribution of each specific research line would have been assigned a place *a priori* might have wreaked of imperialism and intellectual autocracy--*bêtes noires* to the empirical mind. However ecumenical and enterprising were the clarion calls of the pioneers, they probably did not pay serious attention to those deeper epistemological issues which, given the nature of the field, were almost inevitably to arise.[7]

Each of the four major philosophies referred to in today's session could be regarded as having addressed forgotten dimensions of that initial challenge. A popular impression is that positivism has addressed issues of methodological rigor and the verification of statements; that Marxism has pointed to the social interests served by conventional thought and practice; that structuralism has sought

---

4. George Perkins Marsh, *Man and Nature,* ed. David Lowenthal (Cambridge, Massachusetts: Belknap Press, 1964; originally 1864.)

5. Martyn Bowden, "Changes in Land Use in Jefferson County, Nebraska, 1857-1957" (unpublished master's thesis, University of Nebraska, 1959).

6. Gilbert F. White, *Changes in Urban Occupance of Flood Plains in the United States,* University of Chicago Department of Geography Research Paper no. 57 (Chicago: University of Chicago, Department of Geography, 1959); Robert W. Kates, *Hazard and Choice Perception in Flood Plain Management,* University of Chicago Department of Geography Research Paper no. 78 (Chicago: University of Chicago, Department of Geography, 1962).

7. Robert W. Kates and J. Wohlwill, eds., *Journal of Social Issues* 22 (1966); Lowenthal, ed., *Environmental Perception and Behavior.*

to undermine the myth of human subjects as agents of consciousness and action; and that phenomenology has made people aware of their *a priori* lenses, suggesting that they be placed in parentheses, so that reality could speak for itself. All four perspectives begin with a distinct set of assumptions about the nature of being and thought; each, too, bears the stamp of the period and place in which it was first articulated and developed. Transposed to other contexts, to periods and places with different challenges and engrained habits of thought, there are inevitable losses and gains in meaning, inevitable misunderstandings and difficulties in communication. The credibility of their distinct claims to truth can scarcely be understood unless there is an appreciation of the contexts to which they addressed themselves initially, and the circumstances surrounding their transposition from one context to another.

I do not refer simply to the question of translations from German and French, or the purely linguistic aspects of transposing ideas from continental regionalism to Anglo-American empiricism. Far more significant, it seems to me, are issues of cultural value within host and guest worlds, e.g., assumptions about the nature of reality, the role of scholars in society, and the relationships between academic research and problem-solving. As Wittgenstein and others have shown, focus on "translations" may lead one to a thoroughly relativistic position, e.g., "it's all in the game of language," or "it depends on how you define x or y." In other words, the possibility of reaching common denominators of semantics, let alone logic, or the prospects of debate or dialogue among these four schools today, are seriously diminished if the contextually defined values of each perspective are not acknowledged. This holds not only for schools of thought but also maybe for our own individual contributions as well.

Let me illustrate from my own experience. When I returned to the United States in 1970 after some exploration into the everyday worlds of relocated Glasgow families, I was given to understand that if the notion of "social space" were ever to become acceptable, I should develop a mode of operationalizing it, viz., develop a reliable instrument to test the idea and then deliver some useful results, for example, in housing policy.[8] The values of my "host" context, thus, were those of analytical acuity and rigor on the one

---

8. Anne Buttimer, "Social Space and the Planning of Residential Areas," *Environment and Behavior* 4 (1972): 279-318; reprinted in *The Human Experience of Space and Place,* ed. Anne Buttimer and David Seamon (New York: St. Martin's, 1980), pp. 21-54.

hand, and of practical usefulness on the other. Now, I was in an existentialist mood at the time and such technical or pragmatic concerns, however exciting or challenging, could only become meaningful if they could be harmonized with a particular value system. Research "objects" would have to be regarded as "subjects"; the researcher would ideally become a facilitator of heightened awareness among those subjects and also encourage them to assume more discretion and responsibility over their own life situations. There were, in other words, "value" questions intruding all the time on conversations and debates about environmental perception and behavior which could not be handled simply by translations of x's and y's.

No doubt each of today's panelists has had similar experiences of such conversations where people seem to be talking past each other. As long as the other is perceived in the categories of one's own current "ism", is it possible to really hear what that other person is saying? The Anglo-American penchant for giving greater credence to ideas which can be operationalized analytically and can yield practical solutions to problems may be one of the reasons why there is such confusion and misunderstanding among these four schools of thought. Each of them addresses a distinct constellation of intellectual and practical curiosities and bears a distinct image of the world; to judge the validity of one perspective in the categories of another may lead to a kind of "blindman's buff" rather than an exercise in scholarly communication.

Positivism, it is generally felt, had the easiest crossing from Europe to America. It came already mature and well equipped to negotiate with and build on the predominant values of its host world. It proclaimed verifiability of statement and horizons of scientific status for geography on the one hand, as well as prospects for theoretically-grounded contributions to policy-making on the other. For geographers and psychologists in the postwar era, what more welcome news was conceivable for those who had played hero in the defense of freedom and rationality? One can understand the anger aroused by any subsequent migrant perspective which would challenge the hegemony of positivism, especially if the foundations of that perspective were not immediately translatable into technical and pragmatic terms.

But what if one had not experienced World War II or those threats to human freedom so starkly dramatized in the thirties? What if, at another time and place, the problems of poverty, prejudice, alienation and social injustice seemed more urgent? One might have become more concerned about changing the world rather than theorizing about it, as Marx was; more concerned about praxis than with thought. Or, if in the doldrums of the twentieth century, one became more concerned with the quality of life experience rather than the quantity of goods or the "rationality" of things, then issues of personal authenticity and integrity might become interesting, as they were for existentialists. And apart from social relevance altogether, if one's central concerns were epistemological, e.g., clearing the ground for the exercise of the Pure Reason in an era of "-isms" and "-ologies", then the radicalism of *zu den sachen* might hold appeal.

The paper by Golledge and Couclelis extols the virtues of positivism in "setting standards of clarity, consistency, and rigor in the development of argument and conduct of inquiry which are unparalleled in the history of human thought." Acknowledging the varieties of orientation associated with the field, the authors nevertheless identify its primary feature, viz., the analytical method. Then, dismissing one by one most of those tenets which have appeared objectionable or 'unnecessary', they wind up in a position which resembles quite closely American pragmatism. Not only is positivism now to be defined in terms of what positivists do, but "few, if any of the classic positivist tenets now seem to be necessary." What remains after the purge is the *sine qua non* of reductionism in method, and also a set of beliefs and preferences about the nature of reality which fits neatly into the American Dream. There is no place here for the poorly trained, the faint hearted, or the dilettante. The grounds on which the other three schools are dismissed also stem from aesthetic and moral preferences. Marxism is unwelcome because it does not acknowledge "human beings as the primary and most important reality"; structuralism threatens to make the field "social at the cost of ceasing to be human"; and phenomenology is seen as being concerned with "disembodied consciousness hovering over a world of thought and feelings" and "perpetuating the alienating and unnecessary dichotomy between body and soul." The authors' perception of these three schools and the criteria on which they are to be judged reflect cultural values and ideological choices quite as much as their avowed epistemological commitments.

I'm not sure all positivists would be so flexible, but suppose one does take the option outlined here. One still has to confront the issue of whether the epistemological claims of positivism are that reconcilable with those of pragmatism. Can one still expect general theories, laws, and verifiable statements about environmental perception and behavior? Pragmatists claim a contextual explanation of discrete events, and, in varieties of radical empiricism, no claims to scientific law or perennial truth are forwarded: "The truth is only that expedient in the way of our thinking... we have to live today by what truth we get today and be ready tomorrow to call it falsehood."[9] As I understand it, a pragmatist would take each event in terms of its own coordinates of reference in time, space and movement; contingency, spontaneity and unpredictability are cherished values. Now that seems a far cry from the positivist search for general laws about time, space, and behavior. I wonder, therefore, how Golledge and Couclelis reconcile these apparently countervailing claims of pragmatism and positivism when it comes to future development in environmental perception and behavior.

Detective stories can compete any day with good operational models. The title of Douglas Greenberg's paper already launches his theme in an American idiom. In fact, nothing in his preamble should raise anybody's suspicions until that cryptic citation from Marx: "we make our own history, but not of our own free will."[10] One senses that there is more implied here than qualifications about bounded rationality or the resiliency of the analytical method--that something beyond the limits of alleged freedom and dignity holds the key to explaining behavior. Human beings are not to lose their primacy; rather it is "conscious human agency" that gets de-centered. Ontologically speaking, it is *capital* which "must be treated as the directive force propelling history within societies like our own."

There are no doubt many brands of Marxism and structuralism, many reinterpretations of *Das Kapital* and *Tristes Topiques*. It is a brave challenge to bridge the gulf between these two fundamentally contrasting schools of thought. Althusser, Thompson, Giddens and Greenberg notwithstanding, I think the contrast is a more fundamental one and the differences less reconcilable than is acknowledged in Greenberg's paper. Nor do I think one can fully appreciate the whole agency-structure debate without considering the context of France in the 1960s, the non-events of May 1968, and the

---

9. William James, *Pragmatism and Other Essays from the Meaning of Truth* (New York: Meridian, 195), p. 145.

10. The actual text paraphrased here (R. C. Tucker, *The Marx-Engels Reader* [New York: W. W. Norton, 1972]:437) makes a slightly less demanding assertion: "Men make their own history, but they do not make it just as they please; they do not make it under circumstances chosen by themselves, but under circumstances directly found, given and transmitted from the past."

nausea and anger generated afterwards among those who sought to launch a truly revolutionary movement. Anglo commentary on the *nouveaux philosophes* often misses the point about emotional aspects of this literature. The passionate search for alternatives to those idealistic rhetorics of pre-1968 (socialism, Maoism, existentialism and Marxism) was motivated to a large extent by the disillusionment about the "dream and reality" of liberation from structures rather than cool dispassionate reworking of epistemology. Was it not the liberation rather than the definition of man that Marx set as goal when he placed *praxis* instead of consciousness at the heart of his efforts--sentient-imbued praxis anchored in the everyday realities of work and social interaction?[11] In this vein, there was a wide arena for potential interaction with existentialists and others who had read Hegel and appreciated the values of dialectical reasoning.[12] Althusser claims that somewhere between 1845 and 1848, there was an epistemological turn in Marx: his earlier notion of praxis (which in his view still smacked of German idealism) was subordinated to a more scientific theory of history based on distinctions between superstructure and infrastructure as well as the economic determinism of production forces and exchange relations.[13] Thus, he sought to reconcile Marxism and structuralism--a *tour de force,* indeed--for structuralism, in its genesis and *raison d'être,* had set out to erase history, to dispel the myths of human subjectivity, and to replace concern about human agency with observations into an anonymous archipelago of structures. For Levi-Strauss, one of the key founders of structuralism, the issue was "not to constitute but to dissolve man," to eliminate "particular, finite, historical subjectivity."[14] Mind, of course, could survive, but this was a kind of trans-historical objective mind, "unconcerned with the identity of its occasional bearers."[15]

In eliminating concern about diachronic movements in social history, then, and in focusing on the invariant grammars of objective mind, structuralists are evidently prepared to lose touch with the concrete realities and spontaneous accidents of lived

---

11. Tucker, op. cit.

12. Jean-Paul Sartre, *Search for a Method* (New York: Knopf, 1963); Maurice Merleau-Ponty, *The Adventures of the Dialectic* (Evanston, Illinois: Northwestern University Press, 1973); Erich Fromm, *Marx's Concept of Man* (New York: Frederick Ungar, 1961).

13. Leon Althusser, *For Marx* (New York: Pantheon, 1969).

14. Claude Levi-Strauss, *The Savage Mind* (Chicago: University of Chicago Press, 1966), p. 247.

15. Ibid.

experience. In contrast, Marxists are concerned deeply about those concrete actions and interactions of people in history. However, if all history and, in fact, all human science is "ideology" as Marxists also claim, how can there be a scientific study of environmental perception? Strictly speaking, they should regard the kind of research which we have been pursuing over the past fifteen years as serving the false consciousness which permeates buorgeois ideas and techniques. One wonders how and where, then, is it possible to find some standpoint of inquiry from which to purify our own consciousness of its falsity? Does a refuge in structures or structuration processes provide a solution, or does this lead to another as yet unexploited form of determinism?

The real struggle, it seems to me, is between structure and praxis. And, again, it is not always on the logics of respective procedures, or the analytical sharpness of research techniques or claims to truth that these two schools of thought are propounded or confounded. As with other perspectives, Marxism and structuralism have been evaluated according to the aesthetic, moral, and practical challenges of particular contexts. Thompson's critique of Althusser, in Greenberg's reading, seems to have been a bold attempt to keep the frontier spirit alive--human agency must remain an essential assumption--and, incidentally, to take out a claim for sociology and social history in the debate. Giddens tries to give structuralism a more dynamic flavor, to reach further in the direction of time and space analysis but to save those precious cultural values of enterprise, ingenuity, and theoretically-grounded formulae for action.

Greenberg is enthusiastic about the suggestion that structures be regarded as "enabling" as well as "constraining," i.e, that one looks at structuration processes and "situated practices." It strikes me that there may be an analogy here with the turn toward transactional analysis which researchers in environmental perception took a few years ago in order to transcend the ideological impasse between "behaviorists" and "intentionalists." In other words, one again seeks refuge in a "process" orientation: the primary *explicanda* are not the geometry and architecture of structures and forms (be they images, classes, institutions or regions); rather they are the dynamics of systems within which behavior is situated. If my hunch is correct, then what has been accomplished with the "structuration" proposal is a possible solution to the methodological problem, a shift from a correspondence to an operational definition of truth. What remains unresolved, or is still left in parentheses, are those ontological issues on which the paper began, such as free will and determinism.

Edward Relph's paper moves to a different wavelength. Rather than arguing over conceptualizations, he tells a story about his own tastes and preferences and elaborates on the complex task of landscape description. In his skepticism over theories and abstraction he, too, echoes William James and the radical empiricism of early American philosophy. And from here he shows a kinship of spirit with Ruskin and Turner, Goethe and Beethoven, eventually incorporating some of Heidegger's reflections on thought and being. Relph first de-centers the strictly cognitive questions and then returns to them after an excursion into feeling and seeing.

In style and content this paper is deliberately provocative. All forms of explanation, particularly that of the "behaviorists," are lumped together under the label of potential manipulators; social scientists are either naive or pharisaical. The only worthwhile issue is the researcher's own mode of perception and behavior. For those who have come to cherish "clarity of argument" and "inter-subjectively negotiated categories of discourse," this style is no doubt more than a puzzlement: chasms of language and ideology yawn between this paper and the preceding two. If Relph's intention was to shock, then he has probably succeeded. However, given the context, one in which action speaks louder than words, one in which phenomenology and existentialism are to be defined in terms of what its practitioners do, this style may run the risk of misunderstanding and closing doors on further conversation. Relph, of course, decries all labels, but hearers and readers may be unwilling to refrain from labelling. For other students who have seriously tried to understand these schools of thought and to harvest some insight from them on questions of environmental perception and behavior, Relph's paper may be offensive. It could reinforce those stereotypical perceptions which already becloud the prospects of communication and mutual understanding. His paper could well be construed as reinforcing the mythology of Snow's Two Cultures; his argument is uncompromising on the possibility of any general knowledge at all and ultimately pessimistic about progress in science.

What struck me most in Relph's paper was the manner in which it dramatized that whole "deconstruction" ethos of modernist thought since Wittgenstein--a movement which has sought to dismantle successively traditional foundations of knowledge and to move from epistemology to hermeneutics.[16] The hermeneutical challenge implicit here--the prospect of really being able to read the texts of landscape--does seem to hold promise not only in answering geography's need to become aware of reflexivity, but also in contributing to the elucidation of some dilemmas in modern philosophy. After all, when scientists and humanists come together, what does a geographer have to bring to the meeting? One contribution is surely explication of landscape--reading the face of the earth as the texts in which civilizations have inscribed their ongoing dreams and realities. *Zu den sachen* is even a more radical challenge than Relph or Seamon suggest in their papers, for it points to horizons which must remain, by definition, unattainable via human language, viz., letting the things speak for themselves. To the pragmatist's retort, "show me how it works," a phenomenologist might say, "try it yourself." For "working" phenomenologically is essentially the practice of *épochè*--the placing in parentheses of one's taken-for-granted habits of thought and practice. How this approach is to yield fruit "operationally" is still up to the individual thinker.

Words, symbols, language--communicating insight and experience--remain the central dilemma among our various schools of thought on environmental perception and behavior. Relph has personified the individual's stance on the challenge, and Seamon in his paper attempts to identify areas of research in which the phenomenological perspective may be applied. Illustrating these realities in metaphors of landscape painting, classical music, or agrarian dwelling may, however, imply unwillingness to work on the agenda seen as most urgent by the pragmatist, Marxist, or positivist. Is it really possible to let the things around us--the lego landscapes, the interchangeable components of environments and places, the neon-speckled panoramas through which the urban commuter moves--speak for themselves? If one illustrates the phenomenological perspective only with examples from rustic harmony and agrarian dwelling, may one suspect that the *épochè* has not really been accomplished--that the author's own aesthetic and ideological preferences may still impose a selective filtering on the "things themselves?"[17] Grady Clay has made some strides with the use of ordinary language, inventing phrases and works to capture

---

16. Paul Feyerabend, *Knowledge Without Foundations* (Oberling, Ohio: Oberling College Press, 1961); Richard Rorty, *Philosophy and the Mirror of Nature* (Princeton, New Jersey: Princeton University Press, 1979).

17. David Seamon, "Heidegger's Notion of Dwelling and One Concrete Interpretation as Indicated by Hassan Fathy's "Architecture for the Poor," *Geosciences and Man,* 5 (1983), forthcoming.

features of urban landscapes, and Norberg-Schulz has attempted to discern connections between existential and architectural spaces.[18] How do such endeavors relate to those urgent concerns of the pragmatist, Marxist and positivist? Hassan Fathy's architecture may make marvelous sense phenomenologically but evidently not politically.[19] Could it be that one has to consider the possibilities of film, music, art and sculpture--even imaginative mapping--as more appropriate media through which landscape realities may speak for themselves?

To bring such phenomenological possibilities to harvest, however, one must confront another vital issue: whether individual insight and personal readings of landscapes remain idiosyncratic, or whether a sharing of interpretations is eventually possible. The epistemological challenge translates automatically into a social one. Relph sees no problem here, but I suspect that Seamon does. The printed word of this proceedings volume may parade constellations of fascinating monologue but may never lead to mutually enriching dialogue. It is far easier to spell out contrasts among the four traditions above than it is to find potentially common threads among them. The only inkling of common ground seems to be an implicit reverence for the values of pragmatism and the American Dream--not too different from the values evident in the pioneering panel of 1965. The corollary to e *pluribus unum* would, I presume, be a need for democracy of effort with all the checks and balances thereby implied. The operational formula, however, may be the necessary but not sufficient condition for an *agora* or *koinonia* of research effort on environmental perception and behavior.

Whatever the difficulties and potential obstacles, it is indeed heartening to witness the public articulation of views which were virtually unheard of a generation ago. One hopes that such ventilation leads to more than tokenism, and that the juxtaposed profiles may move into a dance of diverse and mutually stimulating participants. The criteria on which directions for future research are to be discerned today can no longer be simply those of technical ingenuity and policy relevance; assumptions about science and practice can no longer remain unexamined. Doors between disciplines other than psychology have opened wide, and academic interaction across the Channel and Atlantic seems to have quickened. Few

---

18. Grady Clay, *Close Up: How to Read the American City* (New York: Praeger, 1973); Christian Norberg-Schulz, *Genius Loci: Toward a Phenomenology of Architecture* (New York: Rizzoli, 1980).

19. Seamon, loc. cit.

critics can any longer deny that scholarly and societal interests are intimately interwoven or that ideas and practices need to be understood contextually. Insofar as our research effort during the past fifteen years has contributed toward these insights, one can feel justifiably gratified. Insofar as responsibility for choices in both the intellectual quality and political economy of research has increased, one feels somewhat baffled, but hopeful.

THE UNIVERSITY OF CHICAGO
DEPARTMENT OF GEOGRAPHY
RESEARCH PAPERS (Lithographed, 6 × 9 inches)

LIST OF TITLES IN PRINT

48. BOXER, BARUCH. *Israeli Shipping and Foreign Trade.* 1957. 162 p.
56. MURPHY, FRANCIS C. *Regulating Flood-Plain Development.* 1958. 216 pp.
62. GINSBURG, NORTON, editor. *Essays on Geography and Economic Development.* 1960. 173 p.
71. GILBERT, EDMUND WILLIAM. *The University Town in England and West Germany.* 1961. 79 p.
72. BOXER, BARUCH. *Ocean Shipping in the Evolution of Hong Kong.* 1961. 108 p.
91. HILL, A. DAVID. *The Changing Landscape of a Mexican Municipio, Villa Las Rosas, Chiapas.* 1964. 121 p.
97. BOWDEN, LEONARD W. *Diffusion of the Decision To Irrigate: Simulation of the Spread of a New Resource Management Practice in the Colorado Northern High Plans.* 1965. 146 pp.
98. KATES, ROBERT W. *Industrial Flood Losses: Damage Estimation in the Lehigh Valley.* 1965. 76 pp.
101. RAY, D. MICHAEL. *Market Potential and Economic Shadow: A Quantitative Analysis of Industrial Location in Southern Ontario.* 1965. 164 p.
102. AHMAD, QAZI. *Indian Cities: Characteristics and Correlates.* 1965. 184 p.
103. BARNUM, H. GARDINER. *Market Centers and Hinterlands in Baden-Württemberg.* 1966. 172 p.
105. SEWELL, W. R. DERRICK, et al. *Human Dimensions of Weather Modification.* 1966. 423 p.
106. SAARINEN, THOMAS FREDERICK. *Perception of the Drought Hazard on the Great Plains.* 1966. 183 p.
107. SOLZMAN, DAVID M. *Waterway Industrial Sites: A Chicago Case Study.* 1967. 138 p.
108. KASPERSON, ROGER E. *The Dodecanese: Diversity and Unity in Island Politics.* 1967. 184 p.
109. LOWENTHAL, DAVID, editor, *Environmental Perception and Behavior.* 1967. 88 p.
112. BOURNE, LARRY S. *Private Redevelopment of the Central City, Spatial Processes of Structural Change in the City of Toronto.* 1967. 199 p.
113. BRUSH, JOHN E., and GAUTHIER, HOWARD L., JR., *Service Centers and Consumer Trips: Studies on the Philadelphia Metropolitan Fringe.* 1968. 182 p.
114. CLARKSON, JAMES D., *The Cultural Ecology of a Chinese Village: Cameron Highlands, Malaysia.* 1968. 174 p.
115. BURTON, IAN, KATES, ROBERT W., and SNEAD, RODMAN E. *The Human Ecology of Coastal Flood Hazard in Megalopolis.* 1968. 196 p.
117. WONG, SHUE TUCK, *Perception of Choice and Factors Affecting Industrial Water Supply Decisions in Northeastern Illinois.* 1968. 93 p.
118. JOHNSON, DOUGLAS L. *The Nature of Nomadism: A Comparative Study of Pastoral Migrations in Southwestern Asia and Northern Africa.* 1969. 200 p.
119. DIENES, LESLIE. *Locational Factors and Locational Developments in the Soviet Chemical Industry.* 1969. 262 p.
120. MIHELIČ, DUŠAN. *The Political Element in the Port Geography of Trieste.* 1969. 104 p.
121. BAUMANN, DUANE D. *The Recreational Use of Domestic Water Supply Reservoirs: Perception and Choice.* 1969. 125 p.
122. LIND, AULIS O. *Coastal Landforms of Cat Island, Bahamas: A Study of Holocene Accretionary Topography and Sea-Level Change.* 1969. 156 p.
123. WHITNEY, JOSEPH B. R. *China: Area, Administration and Nation Building.* 1970. 198 p.
124. EARICKSON, ROBERT. *The Spatial Behavior of Hospital Patients: A Behavioral Approach to Spatial Interaction in Metropolitan Chicago.* 1970. 138 p.
125. DAY, JOHN CHADWICK. *Managing the Lower Rio Grande: An Experience in International River Development.* 1970. 274 p.
126. MaCIVER, IAN. *Urban Water Supply Alternatives: Perception and Choice in the Grand Basin Ontario.* 1970. 178 p.
127. GOHEEN, PETER G. *Victorian Toronto, 1850 to 1900: Pattern and Process of Growth.* 1970. 278 p.
128. GOOD, CHARLES M. *Rural Markets and Trade in East Africa.* 1970. 252 p.
129. MEYER, DAVID R. *Spatial Variation of Black Urban Households.* 1970. 127 p.
130. GLADFELTER, BRUCE G. *Meseta and Campiña Landforms in Central Spain: A Geomorphology of the Alto Henares Basin.* 1971. 204 p.

131. NEILS, ELAINE M. *Reservation to City: Indian Migration and Federal Relocation.* 1971. 198 p.
132. MOLINE, NORMAN T. *Mobility and the Small Town, 1900–1930.* 1971. 169 p.
133. SCHWIND, PAUL J. *Migration and Regional Development in the United States.* 1971. 170 p.
134. PYLE, GERALD F. *Heart Disease, Cancer and Stroke in Chicago: A Geographical Analysis with Facilities, Plans for 1980.* 1971. 292 p.
135. JOHNSON, JAMES F. *Renovated Waste Water: An Alternative Source of Municipal Water Supply in the United States.* 1971. 155 p.
136. BUTZER, KARL W. *Recent History of an Ethiopian Delta: The Omo River and the Level of Lake Rudolf.* 1971. 184 p.
139. MCMANIS, DOUGLAS R. *European Impressions of the New England Coast, 1497–1620.* 1972. 147 p.
140. COHEN, YEHOSHUA S. *Diffusion of an Innovation in an Urban System: The Spread of Planned Regional Shopping Centers in the United States, 1949–1968,* 1972. 136 p.
141. MITCHELL, NORA. *The Indian Hill-Station: Kodaikanal.* 1972. 199 p.
142. PLATT, RUTHERFORD H. *The Open Space Decision Process: Spatial Allocation of Costs and Benefits.* 1972. 189 p.
143. GOLANT, STEPHEN M. *The Residential Location and Spatial Behavior of the Elderly: A Canadian Example.* 1972. 226 p.
144. PANNELL, CLIFTON W. *T'ai-chung, T'ai-wan: Structure and Function.* 1973. 200 p.
145. LANKFORD, PHILIP M. *Regional Incomes in the United States, 1929–1967: Level, Distribution, Stability, and Growth.* 1972. 137 p.
146. FREEMAN, DONALD B. *International Trade, Migration, and Capital Flows: A Quantitative Analysis of Spatial Economic Interaction.* 1973. 201 p.
147. MYERS, SARAH K. *Language Shift Among Migrants to Lima, Peru.* 1973. 203 p.
148. JOHNSON, DOUGLAS L. *Jabal al-Akhdar, Cyrenaica: An Historical Geography of Settlement and Livelihood.* 1973. 240 p.
149. YEUNG, YUE-MAN. *National Development Policy and Urban Transformation in Singapore: A Study of Public Housing and the Marketing System.* 1973. 204 p.
150. HALL, FRED L. *Location Criteria for High Schools: Student Transportation and Racial Integration.* 1973. 156 p.
151. ROSENBERG, TERRY J. *Residence, Employment, and Mobility of Puerto Ricans in New York City.* 1974. 230 p.
152. MIKESELL, MARVIN W., editor. *Geographers Abroad: Essays on the Problems and Prospects of Research in Foreign Areas.* 1973. 296 p.
153. OSBORN, JAMES F. *Area, Development Policy, and the Middle City in Malaysia.* 1974. 291 p.
154. WACHT, WALTER F. *The Domestic Air Transportation Network of the United States.* 1974. 98 p.
155. BERRY, BRIAN J. L., et al. *Land Use, Urban Form and Environmental Quality.* 1974. 440 p.
156. MITCHELL, JAMES K. *Community Response to Coastal Erosion: Individual and Collective Adjustments to Hazard on the Atlantic Shore.* 1974. 209 p.
157. COOK, GILLIAN P. *Spatial Dynamics of Business Growth in the Witwatersrand.* 1975. 144 p.
159. PYLE, GERALD F. et al. *The Spatial Dynamics of Crime.* 1974. 221 p.
160. MEYER, JUDITH W. *Diffusion of an American Montessori Education.* 1975. 97 p.
161. SCHMID, JAMES A. *Urban Vegetation: A Review and Chicago Case Study.* 1975. 266 p.
162. LAMB, RICHARD F. *Metropolitan Impacts on Rural America.* 1975. 196 p.
163. FEDOR, THOMAS STANLEY. *Patterns of Urban Growth in the Russian Empire during the Nineteenth Century.* 1975. 245 p.
164. HARRIS, CHAUNCY D. *Guide to Geographical Bibliographies and Reference Works in Russian or on the Soviet Union.* 1975. 478 p.
165. JONES, DONALD W. *Migration and Urban Unemployment in Dualistic Economic Development.* 1975. 174 p.
166. BEDNARZ, ROBERT S. *The Effect of Air Pollution on Property Value in Chicago.* 1975. 111 p.
167. HANNEMANN, MANFRED. *The Diffusion of the Reformation in Southwestern Germany, 1518–1534.* 1975. 248 p.
168. SUBLETT, MICHAEL D. *Farmers on the Road. Interfarm Migration and the Farming of Noncontiguous Lands in Three Midwestern Townships. 1939–1969.* 1975. 228 pp.
169. STETZER, DONALD FOSTER. *Special Districts in Cook County: Toward a Geography of Local Government.* 1975. 189 pp.
170. EARLE, CARVILLE V. *The Evolution of a Tidewater Settlement System: All Hallow's Parish, Maryland, 1650–1783.* 1975. 249 pp.
171. SPODEK, HOWARD. *Urban-Rural Integration in Regional Development: A Case Study of Saurashtra, India—1800–1960.* 1976. 156 pp.

172. COHEN, YEHOSHUA S. and BERRY, BRIAN J. L. *Spatial Components of Manufacturing Change.* 1975. 272 pp.
173. HAYES, CHARLES R. *The Dispersed City: The Case of Piedmont, North Carolina.* 1976. 169 pp.
174. CARGO, DOUGLAS B. *Solid Wastes: Factors Influencing Generation Rates.* 1977. 112 pp.
175. GILLARD, QUENTIN. *Incomes and Accessibility. Metropolitan Labor Force Participation, Commuting, and Income Differentials in the United States, 1960–1970.* 1977. 140 pp.
176. MORGAN, DAVID J. *Patterns of Population Distribution: A Residential Preference Model and Its Dynamic.* 1978. 216 pp.
177. STOKES, HOUSTON H.; JONES, DONALD W. and NEUBURGER, HUGH M. *Unemployment and Adjustment in the Labor Market: A Comparison between the Regional and National Responses.* 1975. 135 pp.
179. HARRIS, CHAUNCY D. *Bibliography of Geography. Part I. Introduction to General Aids.* 1976. 288 pp.
180. CARR, CLAUDIA J. *Pastoralism in Crisis. The Dasanetch and their Ethiopian Lands.* 1977. 339 pp.
181. GOODWIN, GARY C. *Cherokees in Transition: A Study of Changing Culture and Environment Prior to 1775.* 1977. 221 pp.
182. KNIGHT, DAVID B. *A Capital for Canada: Conflict and Compromise in the Nineteenth Century.* 1977. 359 pp.
183. HAIGH, MARTIN J. *The Evolution of Slopes on Artificial Landforms: Blaenavon, Gwent.* 1978. 311 pp.
184. FINK, L. DEE. *Listening to the Learner. An Exploratory Study of Personal Meaning in College Geography Courses.* 1977. 200 pp.
185. HELGREN, DAVID M. *Rivers of Diamonds: An Alluvial History of the Lower Vaal Basin.* 1979. 399 pp.
186. BUTZER, KARL W., editor. *Dimensions of Human Geography: Essays on Some Familiar and Neglected Themes.* 1978. 201 pp.
187. MITSUHASHI, SETSUKO. *Japanese Commodity Flows.* 1978. 185 pp.
188. CARIS, SUSAN L. *Community Attitudes toward Pollution.* 1978. 226 pp.
189. REES, PHILIP M. *Residential Patterns in American Cities, 1960.* 1979. 424 pp.
190. KANNE, EDWARD A. *Fresh Food for Nicosia.* 1979. 116 pp.
191. WIXMAN, RONALD. *Language Aspects of Ethnic Patterns and Processes in the North Caucasus.* 1980. 224 pp.
192. KIRCHNER, JOHN A. *Sugar and Seasonal Labor Migration: The Case of Tucumán, Argentina.* 1980. 158 pp.
193. HARRIS, CHAUNCY D. and FELLMANN, JEROME D. *International List of Geographical Serials, Third Edition, 1980.* 1980. 457 p.
194. HARRIS, CHAUNCY D. *Annotated World List of Selected Current Geographical Serials, Fourth, Edition. 1980.* 1980. 165 p.
195. LEUNG, CHI-KEUNG. *China: Railway Patterns and National Goals.* 1980. 235 p.
196. LEUNG, CHI-KEUNG and GINSBURG, NORTON S., eds. *China: Urbanization and National Development.* 1980. 280 p.
197. DAICHES, SOL. *People in Distress: A Geographical Perspective on Psychological Well-being.* 1981, 199 p.
198. JOHNSON, JOSEPH T. *Location and Trade Theory: Industrial Location, Comparative Advantage, and the Geographic Pattern of Production in the United States.* 1981. 107 p.
199-200. STEVENSON, ARTHUR J. *The New York-Newark Air Freight System.* 1982. 440 p. (Double number, price: $16.00)
201. LICATE, JACK A. *Creation of a Mexican Landscape: Territorial Organization and Settlement in the Eastern Puebla Basin, 1520–1605.* 1981. 143 p.
202. RUDZITIS, GUNDARS. *Residential Location Determinants of the Older Population.* 1982. 117 p.
203. LIANG, ERNEST P. *China: Railways and Agricultural Development, 1875–1935.* 1982. 186 p.
204. DAHMANN, DONALD C. *Locals and Cosmopolitans: Patterns of Spatial Mobility during the Transition from Youth to Early Adulthood.* 1982. 146 p.
205. FOOTE, KENNETH E. *Color in Public Spaces: Toward a Communication-Based Theory of the Urban Built Environment.* 1983. 153 p.
206. HARRIS, CHAUNCY D. *Bibliography of Geography. Part II: Regional. Vol. 1. The United States of America.* 1984. 178 p.
207-208. WHEATLEY, PAUL. *Nāgara and Commandery: Origins of the Southeast Asian Urban Traditions.* 1983. 473 p. (Double number, price: $16.00)

209. SAARINEN, THOMAS F.; SEAMON, DAVID; and SELL, JAMES L., eds. *Environmental Perception and Behavior: An Inventory and Prospect.* 1984. 263 p.
210. WESCOAT, JAMES L., JR. *Integrated Water Development: Water Use and Conservation Practice in Western Colorado.* 1984. 239 p.